# Gender Politics in the Asia-Pacific Region

Amidst the unevenness and unpredictability of change in the Asia-Pacific region, women's lives are being transformed. This volume takes up the challenge of exploring the ways in which women are active players, collaborators, participants, leaders and resistors in the politics of change in the region.

The contributors argue that 'gender' matters and continues to make a difference in the midst of change, even as it is intertwined with questions of tradition, generation, ethnicity and nationalism. Drawing on current dialogue among feminism, cultural politics and geography, the book focuses on women's agencies and activisms, insisting on women's strategic conduct in constructing their own multiple identities and navigation of their life paths.

The editors focus attention on the politics of gender as a mobilising centre for identities, and the ways in which individualised identity politics may be linked to larger collective emancipatory projects based on shared interests, practical needs or common threats. Collectively, the chapters illustrate the complexity of women's strategies, the diversity of sites for action, and the flexibility of their alliances as they carve out niches for themselves in what are still largely patriarchal worlds.

This book will be of vital interest to scholars in a range of subjects, including gender studies, human geography, women's studies, Asian studies, sociology and anthropology.

**Brenda S.A. Yeoh** is an Associate Professor in the Department of Geography, National University of Singapore. Her research foci include the politics of space in colonial and post-colonial cities, and gender, migration and transnational communities.
**Peggy Teo** is an Associate Professor in the Department of Geography, National University of Singapore. She has research interests in tourism and social gerontological issues.
**Shirlena Huang** is an Associate Professor in the Department of Geography, National University of Singapore. Her research interests focus mainly on gender issues (with specific interest in transnational labour migration in the Asia-Pacific) as well as urban conservation and heritage.

# Routledge International Studies of Women and Place

Series editors: Janet Henshall Momsen and Janice Monk

**1 Gender, Migration and Domestic Service**
Edited by Janet Henshall Momsen

**2 Gender Politics in the Asia-Pacific Region**
Edited by Brenda S.A. Yeoh, Peggy Teo and Shirlena Huang

**3 Geographies of Women's Health**
Place, Diversity and Difference
Edited by Isabel Dyck, Nancy Davis Lewis and Sara McLafferty

**4 Gender, Migration and the Dual Career Household**
Irene Hardill

## Also available from Routledge:

**Full Circles**
Geographies of Women over the Life Course
Edited by Cindi Katz and Janet Monk

**'Viva'**
Women and Popular Protest in Latin America
Edited by Sarah A. Radcliffe and Sallie Westwood

**Different Places, Different Voices**
Gender and Development in Africa, Asia and Latin America
Edited by Janet Momsen and Vivian Kinnaird

**Servicing the Middle Classes**
Class, Gender and Waged Domestic Labour in Contemporary Britain
Nicky Gregson and Michelle Lowe

**Women's Voices from the Rainforest**
Janet Gabriel Townsend

**Gender, Work and Space**
Susan Hanson and Geraldine Pratt

**Women and the Israeli Occupation**
Edited by Tamar Mayer

**Feminism / Postmodernism / Development**
Edited by Marianne H. Marchand and Jane L. Parpart

**Women of the European Union**
The Politics of Work and Daily Life
Edited by Janice Monk and Maria Dolors Garcia-Raomon

**Who Will Mind the Baby?**
Geographies of Childcare and Working Mothers
Edited by Kim England

**Feminist Political Ecology**
Global Issues and Local Experience
Edited by Dianne Rocheleau, Esther Wangari and Barbara Thomas-Slayter

**Women Divided**
Gender, Religion and Politics in Northern Ireland
Rosemary Sales

**Women's Lifeworlds**
Women's Narratives on Shaping their Realities
Edited by Edith Sizoo

**Gender, Planning and Human Rights**
Edited by Tovi Fenster

**Gender, Ethnicity and Place**
Women and Identity in Guyana
Linda Peake and D. Alissa Trotz

# Gender Politics in the Asia-Pacific Region

Edited by Brenda S.A. Yeoh,
Peggy Teo and Shirlena Huang

Routledge
Taylor & Francis Group

LONDON AND NEW YORK

First published 2002
by Routledge

2 Park Square, Milton Park, Abingdon, Oxfordshire OX14 4RN

Simultaneously published in the USA and Canada
by Routledge

711 Third Avenue, New York, NY 10017
First issued in paperback 2014

*Routledge is an imprint of the Taylor and Francis Group, an informa business*

Typeset in Baskerville by Taylor and Francis Books Ltd

*British Library Cataloguing in Publication Data*
A catalogue record for this book is available from the British Library

*Library of Congress Cataloging in Publication Data*
A catalogue record for this book has been requested

ISBN 978-0-415-20660-0 (hbk)
ISBN 978-0-415-69534-3 (pbk)

# Contents

# Illustrations

## Tables

## Figure

# Contributors

**Rebecca Elmhirst** is Senior Lecturer in Human Geography at the University of Brighton, UK. Her research is focused on gender, migration and ethnic relations, particularly in Southern Sumatra, Indonesia. Her most recent work is concerned with resource politics in transmigration resettlement areas, and with labour politics among female migrant workers in Indonesia.

**Rita S. Gallin** is Professor of Sociology and Women's Studies at Michigan State University, USA. Her research interests include women and change in Taiwan and global transformation. She has published extensively on her longitudinal field research in rural Taiwan and is the founding editor of the *Working Papers on Women and International Development* (Michigan State University) and the *Women and International Development Annual* (Westview Press).

**Tatjana Haque** completed her PhD in Geography at University College London, UK. She is interested in applied research in South Asia (particularly Bangladesh) with focus on women's participation in local governance, training in non-traditional skills, empowerment and political leadership. She has recently been appointed Country Representative and Programme Manager for Christian Aid's field office in Bangladesh, and has authored various publications on gender identities in Bangladesh.

**Shirlena Huang** is an Associate Professor at the Department of Geography, National University of Singapore. Her research interests focus mainly on gender issues (with specific interest in transnational labour migration in the Asia-Pacific) as well as urban conservation and heritage. Her publications include articles in *International Migration Review*, *Women's Studies International Forum*, *Urban Studies* and *Geoforum*.

**Lisa Law** is Assistant Professor, Department of Geography, National University of Singapore. She is a social and cultural geographer interested in issues of migration and identity. Her research interests include the Filipino diaspora in the Asian region, particularly the cultural politics of activism around issues of labour migration to Asia's tiger economies. Lisa is interested in issues of gender, place, migration and identity, and how they become entwined with Filipino women's experiences abroad.

**Jacqueline Leckie** is Senior Lecturer in Social Anthropology, University of Otago, New Zealand. She has published on gender, ethnicity, migration and work in the South Pacific, particularly Fiji. This includes two books, *To Labour with the State* (Dunedin: University of Otago Press, 1997) and the co-edited *Labour in the South Pacific* (Townsville: James Cook University of North Queensland, 1990). Currently she is writing a history of the construction and management of 'madness' in Fiji.

**Vera Mackie** is Foundation Professor of Japanese Studies at Curtin University of Technology in Western Australia. Major publications include *Creating Socialist Women in Japan: Gender, Labour and Activism, 1900–1937* (Cambridge: Cambridge University Press, 1997); *Human Rights and Gender Politics: Asia-Pacific Perspectives* (London: Routledge, 2000, co-edited with Anne Marie Hilsdon, Martha Macintyre and Maila Stivens); *Relationships: Japan and Australia, 1870s–1950s* (Melbourne: History Department Monograph Series, 2001, co-edited with Paul Jones); and articles in such journals as *Australian Feminist Studies, East Asian History, Hecate, International Feminist Journal of Politics, Intersections, Japanese Studies, New Left Review* and *Women's Studies International Forum*.

**Ruth Panelli** is Senior Lecturer, Department of Geography, University of Otago, New Zealand. She is a social geographer specialising in critical and discursive studies of rural society. These interests have led to investigations of gender, youth, health, alternative agriculture and community change in Australian and New Zealand contexts.

**Lily Phua** completed her Master's degree in Geography at the National University of Singapore. Her work has focused mainly on the gendering of space, drawing material from the conduct of everyday life in Singapore.

**Nobue Suzuki** is a PhD candidate in the Department of Anthropology at the University of Hawai'i, Mānoa. She is completing her dissertation, entitled 'Battlefields of affection: gender, global desires and the politics of intimacy in Filipina–Japanese transnational marriages'. She has published several papers in journals such as *Women's Studies International Forum* and *U.S.–Japan Women's Journal English Supplement*, and with James E. Roberson is co-editing a book entitled *Men and Masculinities in Contemporary Japan: Dislocating the Salaryman Doxa*.

**Peggy Teo** is an Associate Professor in the Department of Geography, National University of Singapore. She has research interests in tourism and social gerontological issues. Her works include *Interconnected Worlds: Tourism in Southeast Asia* (Oxford: Pergamon, 2001, co-edited with T.C. Chang and K.C. Ho) and several journal articles on social gerontological issues, some pertaining to women, in *Aging and Society, Journal of Aging Studies, Urban Studies, Geographical Review, International Journal of Population Geography, Journal of Cross-Cultural Gerontology* and *Woman's Studies International Forum*.

**G.G. Weix** is Associate Professor, Department of Anthropology, and Co-Director of the Women's Studies Program at the University of Montana, USA. She has published in the areas of women and gender, political economy and households, Islam and language in Indonesia. She has embarked on a new research project on global English in Southeast Asia.

**Brenda S.A. Yeoh** is Associate Professor, Department of Geography, National University of Singapore. Her research foci include the politics of space in colonial and post-colonial cities, and gender, migration and transnational communities. She has published over fifty scholarly journal papers and is author/editor of a number of books, including *Contesting Space: Power Relations and the Urban Built Environment in Colonial Singapore* (Oxford: Oxford University Press, 1996); *Singapore: A Developmental City State* (Chichester: John Wiley, 1997, with Martin Perry and Lily Kong), and *Gender and Migration* (Cheltenham: Edward Elgar, 2000, with Katie Willis).

# Preface

Jonathan Boyarin (1994: vii) describes the coming together of his collective volume *Remapping Memory: The Politics of TimeSpace* (Minneapolis: University of Minnesota Press) as 'the result of contingent connections between specialists in widely divergent areas who took advantage of an opportunity to articulate a common agenda'. We can think of no better way to describe the development of this book, except to add that the connective tissues which bind this volume together are constituted not only by a range of specialisms, but also by the contingent and consultative ties of friendship and collegiality, focused in our home department at the National University of Singapore as well as stretched across several continents. Help and encouragement have sprung from many quarters and we would like to take this opportunity to acknowledge our debts.

The book grew out of an August 1997 international conference held in Singapore entitled *Women in the Asia-Pacific Region: Persons, Powers and Politics*, organised by the Department of Geography, the Southeast Asian Studies Programme and the Centre for Advanced Studies, all at the National University of Singapore. We are particularly indebted to Associate Professor Teo Siew Eng, the former head of the Department of Geography, as well as Associate Professor Victor Savage, the present head, for their commitment of resources and encouraging support. Without the support of the Southeast Asian Studies Programme and the Centre for Advanced Studies, we would have had more difficulty advancing the multidisciplinary reach of the conference and, subsequently, the book. We are also very grateful to the International Geographical Union's Commission on Gender and Geography and particularly its former chairperson Ruth Fincher for her stalwart support in promoting the reach of the conference. In all this, we were aided by an organising committee including Lysa Hong, Jessie Poon, T.C. Chang and Henry Yeung, who all worked tirelessly not only in pulling together the myriad threads which go into the making of a conference but also producing what still stands as the most visible item on our bookshelves today – a 743-page tome of conference papers in various shades of pink and fuchsia!

The road from the initial engagement of ideas at the conference to the stitching together of the book was a long, drawn-out process. With a small nucleus of papers from the conference in hand and a specific brief as to what the book should principally address, we sought a range of contributions from a wide number of places which we assessed, selected, returned for revisions, and finally edited. We are grateful for the encouragement of the series editors Janet Momsen and Janice Monk, as well as the practical help of a number of assistants including Mariam Ali, Angela Yak, Chang Siew Ngoh, Theresa Wong and the support staff at the Centre for Advanced Studies.

In setting a 'common agenda' for this volume, we hope not only to illustrate the rich interplay of gender politics and tease out possible epistemological and practical distinctions, but also, where possible, to give 'voice' to women across the Asia-Pacific region. We (and the contributors too) are well aware of the difficulties involved in such an enterprise, and the easy slippage into mere rhetoric. There have been sufficient warnings in the literature that the very act of representing them and inscribing their 'voice' may rob women of their agency. While we have no pat solution to this so-called 'crisis of representation', we have given emphasis in this collection to papers where there is a certain degree of insideness and immediacy, as well as respect for the subjects' agency, even when authors necessarily 'speak for' those whom they research. Even as we work towards a deeper theoretical understanding of the possibilities and limits of gender politics from the perspectives of women in the region, we want to do so without obfuscating a sense of women's subjectivities and desires as they go about the business of everyday living.

It is our hope that this volume represents a small step in advancing the edge of scholarship on the variegated nature of gender politics in the Asia-Pacific region. It is our conviction that if this collaborative venture should indeed succeed, its claim to success must rest on the labours of the large number of women and a few men (we did not intentionally set out to exclude them!) – series editors, conference organisers, authors, interviewees and questionnaire participants, and research assistants – who have generously given this book a priority and a place amidst the demands and struggles of everyday life.

Brenda S.A. Yeoh, Peggy Teo and Shirlena Huang
National University of Singapore

# 1 Introduction

## Women's agencies and activisms in the Asia-Pacific region

*Brenda S.A. Yeoh, Peggy Teo and Shirlena Huang*

## Introduction

The anticipated dawn of what has been termed the Pacific Century saw a spate of literature (for example, Das 1996; Robison and Goodman 1996; Sen and Stivens 1998) on the Asia-Pacific region focusing on the 'new affluence' and runaway success of East and Southeast Asian 'miracle' economies. This was suddenly brought up short as the region floundered in crisis in the closing years of the old century. A volume such as Krishna Sen's and Maila Stivens' (1998) *Gender and Power in Affluent Asia* began asking interesting and innovative questions around the central theme of gender relations surrounding 'consumption' in an era of affluence in the pursuit of modernity and global futures in Asia. While these themes continue to be relevant, attention has now veered towards dealing with the depth and crisis of change in the region. Some of this literature has begun to examine the differential role and experiences of men and women in sustaining the reproduction of global capitalism and economic development in the midst of change. Beyond giving attention to women cast as workers, there has been interest in women's life experiences past the arena of paid work, as well as the broad discourses of power in which women are inserted, formed and reproduced as female subjects within the uneven web of globalisation (Ram and Jolly 1998; Edwards and Roces 2000; Hilsdon *et al.* 2000; Huang *et al.* 2000; Wille and Passl 2001).

Amidst the unevenness and unpredictability of change, women's lives are being transformed – even as they resist or inflect creeping as well as sweeping change – as they become threaded into the intersecting spaces between globalising time–space compression on the one hand, and the particularities of localisms on the other (as well as the multiple liminal spaces 'betwixt and between'). Thus the subject of this volume is to grapple with the multiple sites women occupy (and from which they are sometimes displaced) within and at the overlapping borderlands among the arenas of industrial society, the state, civil society, community circles and the home.

This volume takes up the challenge of exploring the ways women are active players – not truants but collaborators, participants, leaders and resistors – in the politics of change in the region. Drawing on the current dialogue among feminism, cultural politics and geography, the book focuses on *agencies* and *activisms*, insisting on women's strategic conduct in constructing their own (multiple) identities and navigating their (and often their families') life paths, though not always under conditions of their own choosing. Indeed, more has been written on how structural forces such as global capital, the state or some other institutional form act to define gender identities, construct gender relations and impact on different groups of women, than the converse effects of women's agencies and activisms in shaping institutions and structures and altering gender identities and relations. While recognising the inescapable intertwining of structure and agency, the present volume gives weight to the latter in valorising women's strategies as played out under specific conditions of social materiality, from the little tactics of the habitat to the activisms of organised groups and mass mobilisation on the streets.

The volume provides evidence from a spectrum of localities in the Asia-Pacific region to counter stereotypical discourses which invariably portray women of the region either as silent, domesticised housewives cloistered in the private sphere, or eroticised, exoticised objects of male desire. Constantly framed as 'the other' within relations of dominance and dependence, women's capacity and potential to make a difference and their roles in forging alternative modernities within a globalising whole have not been adequately interrogated. A central concern in this volume is then to endow women as subjects with agency, not simply depict them as passive victims of patriarchy and capitalism, or 'vehicles for the realization of transcending systems or projects' (Randall 1998: 185–6).

## Gender politics and resistances

While recognising that it is no longer sufficient to theorise power relations as the expression of any one singular dimension of oppression and that, instead, social inequalities along axes of race, class, gender and citizenship rights are mutually reinforcing, this book focuses attention on the politics of gender as a mobilising centre for identities, and the ways in which, and under what circumstances, individualised identity politics woven into the fabric of the everyday may be linked to larger collective emancipatory projects based on shared interests, practical needs or common threats. By drawing together the politics of gender and group identifications (whether at the level of the family/household, the nation-state or global capitalism) and the ways they engage the rapidly changing material conditions of the biosphere in which women live, this book provides a different angle to 'seeing' and 'understanding' emerging socio-political power relations in the region. By 'politics', then, we include a wide range of activities undertaken by women 'which fall outside the boundaries of conventional politics and therefore not usually deemed to be "political"' (Waylen 1998: 1).

Stivens (2000: 24) argues that feminist attempts to relocate gender politics in the region 'appear to have homed in on human rights' as claims for women's rights as human rights have been facilitated by 'spectacular growth of a global feminist public'. Our conclusions in this volume are more tentative in nature: while some struggles and projects have clear engagements with the global human rights framework, there are myriad others which emerge in spaces somewhat disconnected to, or dislocated from, the 'global' or even 'public' platform. These fragmentary, less-than-completely articulated, and possibly unintended, struggles written into the interstitial spaces of everyday life should not be dismissed. Given that there are multiple oppressions at work in women's lives at different scales, we argue that emancipatory politics can rely on no one single, universal formula but draw on multiple identifications and diverse strategies, sometimes working the ground 'locally', sometimes collapsing the personal and the political in opposing an exclusionary nation or the discriminatory practices of the state, and sometimes by drawing on transnational or global frameworks or discourses. By demonstrating the possibilities, and the difficulties, involved in these variegated enterprises with different spatial reaches, this volume provides further grist to the mill to debunk essentialist notions of and transhistorical claims about gender, and instead argues for the need for 'situated' knowledge and contextualised evaluations in unravelling gender relations in the region.

As has been argued by others, the domain of 'resistance' and 'politics' must be expanded beyond 'heroic acts by heroic people or heroic organisations' (Thrift 1997: 125) (without detracting from the power and poetics of such acts or suggesting that they are scissored out of the fabric of everyday contexts) and reconfigured to include resistant postures, ploys, tactics and strategies woven into 'the practice of everyday life' (de Certeau 1984). Scott's (1985) work, in particular, was highly influential in valorising 'weapons of the weak' as 'everyday forms of resistance', but also attracted the attention of critics like Abu-Lughod (1990: 42) who cautioned against the 'tendency to romanticize resistance, to read all forms of resistance as signs of the ineffectiveness of systems of power and the resilience and creativity of the human spirit in its refusal to be dominated'. While we need to guard against trivialising women's 'resistance' by discerning it in all situations everywhere, removing it from its macropolitical status allows us to appreciate the fluid, unstable nature of power relations. This creates the conceptual and creative space to rewrite the everyday world that women inhabit on its own terms. As Vidler (1978: 28) points out (albeit in a different context), it is often '[b]etween submission to the intolerable and outraged revolt against it' that ordinary people 'somehow defined a human existence within the walls and along the passage of their streets'.

Routledge (1997: 69) further reminds us of the variegated nature of resistance:

> Resistances may be interpreted as fluid processes whose emergence and dissolution cannot be fixed as points in time [or space]. ... [They are] rhizomatic multiplicities of interactions, relations, and acts of becoming ... Any resistance synthesizes a multiplicity of elements and relations without effacing their heterogeneity or hindering their potential for future rearrangement. As rhizomatic practices, resistances take diverse forms, they move in different dimensions, they create unexpected networks, connections, and possibilities. They may invent new trajectories and forms of existence, articulate alternative futures and possibilities, create autonomous zones as a strategy against particular dominating power relations.

Women's 'politics' and strategies of resistance may thus be 'assembled out of the materials and practices of everyday life' (Routledge 1997: 69). Thinking of resistance as 'rhizomatic practices' – the metaphor appropriately insists on a certain 'grounding' of such practices and at the same time conceives of resistance as sprouting both 'above' and 'below ground' – points to the contingent nature of power. At the same time, it allows us to transcend the dichotomy between treating resistant spaces as purely autonomous, 'uncolonised' spaces exterior to or dislocated from the spatial parameters of domination, or as purely 'underside' spaces of social life confined to and reacting against authorised spaces of domination in a 'strategic' fashion where 'each offensive from one side serves as leverage for a counter-offensive from the other' (Foucault 1980: 163–4). Treating resistance as 'rhizomatic' emphasises its creative and elusive nature, as a subjectivity which is ' "polyphonic", plural, working in many discursive registers, many spaces, many times' (Thrift 1997: 135). It is thus important not only to recognise that 'politics' take on many different hues and inhabit different spaces, but also to underscore the way they connect, collide, diverge, transmute, sometimes in unexpected ways, and often moving 'in' and 'out' of spaces of domination.

## Body politics

Geographers in particular have insisted on the value of a spatial understanding of 'politics' and 'resistance', and in grounding geographic metaphors of space, place and positionality (currently all the rage in cultural studies) in situated practices and local contexts. In considering the 'site' of women's agency – that is, the spaces from which women act – many have argued that it is from marginal spaces and liminal interstices that women find multiple resourceful ways to exercise and express agency. Various authors have written about these spaces – the 'space of radical openness' and the 'profound edge' (hooks 1990); the 'margins', 'periphery' and 'underclass of formal power structures' (Hays-Mitchell 1995); 'counter-spaces' (Lefebvre 1991) – and drawn on metaphors of spatiality to demonstrate how women struggle to 'create, conserve and re-create political spaces' (Keith and Pile 1993).

Spaces at the margins are hence not only produced by dominant groups intent on securing conceptual or instrumental control, but also simultaneously drawn upon by subordinate groups resisting exclusionary definitions or tactics and advancing their own claims. They are 'battleground[s] within which and around which conflicting socio-ecological forces of valuation and representation are perpetually at play' (Harvey 2000: 116). The most 'irreducible locus for the determination of all values, meanings, and significations' within these marginal spaces is the human body, 'the measure of all things' (Harvey 2000: 97–8). Not only are different kinds of bodies produced, both materially and representationally, by different processes, and in the course of it all marked by class, race, gender and other distinctions – these bodies are not 'passive products of external processes' – but 'active and transformative in relation to the processes that produce, sustain, and dissolve [them]' (Harvey 2000: 99).

The politics of the gendered body is given central focus in a number of chapters in this volume. Tatjana Haque observes that in contrast to the multiple ways in which 'the body' is analysed in western discourses – she lists 'healthy bodies', 'sexed bodies', 'technobodies', 'virtual bodies', 'third bodies', 'cyborg bodies', 'hybrid bodies' and 'raced bodies' – the mainstream development literature focusing on the non-western world tends to view women's bodies in rather circumscribed ways, either as human resources for productive purposes or targets for population control. Margaret Jolly (1998a: 3, cited in Leckie, this volume) adds that the 'maternal subject position', and by implication the reproductive body, has been given more attention by Asia-Pacific feminists compared to their western counterparts 'often to distinguish themselves from what are perceived to be anti-family tendencies in western feminism or as part of anti-colonial or nationalist movements'. Even within these parameters, however, it is clear that women's bodies are not necessarily passive or docile objects, but also conduits which enable transgression and resistance (Callard 1998: 387).

This is demonstrated in Lily Phua and Brenda Yeoh's chapter, which highlights the salience of the pregnant body, a bodily form which only women assume and, as such, represents a fraught terrain criss-crossed by political strategies on the one hand claiming complete gender equality and neutrality (and hence risking incorporation into a men's world as 'lesser' beings) and, on the other, demanding recognition that women's bodies are different (and hence risking relegation to a different sphere from men's). For example, while some Singaporean Chinese women insist that their pregnant bodies are not 'weak' or 'sick' but 'normal' in order to legitimise continuing their daily routines, others draw on the power that the embodiment of procreational ability confers as a means to elevate their status and wrest a number of 'gains' (for example, special treatment from usually more powerful others, or legitimation of 'irrational' demands such as food cravings). The politics of sameness *vis-à-vis* the politics of difference is being played out in the context of a regulatory regime in Singapore which simultaneously works a number of discursive registers – procreation as nation-building, medicalisation of pregnancy and the legitimacy of the medical gaze, and cultural understandings of the pregnant body as the embodied continuity of male lineage – in order to produce compliant bodies. In countering the web spun out of these discourses around the pregnant body, Singaporean Chinese women show themselves capable of both discursive negotiations – legitimising their own positionings by drawing on counter-discourses (for example, western medical advice as a means to oppose traditional medical prognosis, and vice versa) or exploiting contradictions between competing discourses (for example, the state's 180-degree turnaround from an anti- to a pro-natalist policy) – as well as 'little' tactics of avoidance and non-compliance.

As a terrain of control and transgression, the salience of the body goes beyond issues of reproduction, as several other contributors show. Using the specific case study of Bangladeshi women involved in the Gonoshasthaya Kendra (translated as The People's Health Centre), a non-government organisation (NGO) with a holistic approach to health care and rural development,

Haque shows how 'women's bodily landscapes' and 'bodily conduct' are transformed as they 'practise empowerment through their bodies'. Embodied empowerment is a process which occurs as women 'act and are acted upon', 'see and are seen', 'speak and are spoken to'. It is experienced in and through the way women change their body language, behaviour and presentation of self as a result of their involvement in Gonoshasthaya Kendra: for example, in presenting their bodies in front of unrelated men, or moving about in public spaces with an air of assurance. From the 'landscape' of the individual body, Haque goes on to show that physical solidarity and visibility conferred by the coming together of many women's bodies in street marches have the inherent potential of transforming not just the public landscape but, in turn, local power relations.

In highlighting the centrality of the body as 'a site at which all strategies of control and resistance are registered' (Ong 1991: 307, cited in Weix, this volume), G.G. Weix's chapter unravels the discursive processes and mythologising work transforming a rape and murder victim, Marsinah, a young Indonesian female worker and trade union activist, into 'a shining symbol for workers' rights'. The rape and torture of a violated female body became 'a sign of defeat for labour activism' as well as a galvanising moment for collective action, while an obsession with repeated autopsies performed on the body 'mark[ed] this death as a [continuing] source of anxiety'. Marsinah's 'face', from realistic images to surrealist distortions, was relentlessly reproduced in both material and metaphorical form on magazine covers and as artistic renditions for exhibition. Her gravesite in East Java has been elevated to an honoured place of pilgrimage for young women workers who continued to 'engage her in conversation'. As Weix points out, the Marsinah case resonates with contemporary audiences because 'violence against workers could be condensed in the figure of a single woman violated and left for dead'. In death as in life, the body continues to be drawn upon in the politics of representation and the struggle over cultural meaning.

## Identity politics

As David Harvey (2000: 118–19) explains, the body cannot be construed as the locus of political action without a notion of what concepts such as 'personhood', 'a sense of self' and 'identity' mean. We follow Harvey (2000: 119) in arguing that 'identity' is relational and socially constructed and that the politics of identity constitutes an important mapping of the basic contours of politics and struggle within the social body – the aims, flashpoints and effects – for 'the assumptions that are made about how people are constituted have profound effects not only on the kind of radical politics people can be expected to make but also on the kind of effects that can ultimately be hoped for through political action' (Keith and Pile 1993: 34). As such, identity politics is not 'some sort of surface froth that floats around on top of more important social processes, but something that strikes deep into our ability to transform the social world into concrete knowledges' (Keith and Pile 1993: 31).

The focus on 'politics' puts the emphasis on the contested 'social processes whereby people articulate, assert, challenge, suppress, realign, and co-opt varying hierarchies of identity', often through claims and counter-claims about 'self' and the 'other' (Dickey and Adams 2000: 10). As Appadurai (1996: 14) notes, not all identity claims are motivated by the pursuit of economic, political, or emotional gains as 'the mobilization of markers of group difference may itself be part of a contestation of values *about* difference, as distinct from the consequences of difference for wealth, security or power'. For example, following Aihwa Ong (1991: 306, cited in Weix, this volume), Weix argues that the politics of representation around Marsinah and 'her unquiet death' had less to do with class interests than a 'cultural struggle' on the part of young women workers striving for 'a politics of social memory in which they are actors as well'. In a different context, Ruth Panelli's chapter shows that in order to resist stereotypical images of 'farm women' being automatically identified as 'farmers' wives', participants in the Women in Agriculture movement in Australia explicitly make strategic identity choices in articulating 'ensembles' of subject positions as, in one instance, 'a farmer, a mother, the acting farm accountant, the secretary of a local farming organisation, the member of an industry co-operative, and the committee member of a regional environmental body'.

Recent feminist theory in search of alternative and emancipatory accounts of human subjectivity has highlighted how women are active agents in negotiating and deploying their own identities not only for strategic purposes in resisting or challenging aspects of patriarchy, capitalism, technology and even feminism, but also in order to valorise 'difference'. While some argue that 'identity politics' are individualistic and inward-looking and have come to replace political struggles aimed at social change with a degraded arena of politics where only the personal is deemed legitimate, others see identity politics as indispensable (Yuval-Davis 1994). Far from a withdrawal from politics, identity politics locate 'grids of power which are horizontal and not just vertical', and recognise the diversity of identities and circumstances, and the complexity and contradiction inherent in many struggles (Rosalind Brunt cited in Yuval Davis 1994: 420). As Panelli shows, it is by accommodating multiple identities and drawing on their creative tensions that the Women in Agriculture movement successfully mobilises across great diversity to effect considerable change at different scales. At the same time, identity 'reconstructions' at the personal level are also important evidence of women's ability to reposition themselves socially and economically, as aptly illustrated in the case of a woman in Panelli's chapter who speaks of going to a gathering organised by the movement as 'a farmer's wife' but coming back 'a farmer'. In surveying the field of gender politics in post-coup Fiji, however, Jacqueline Leckie cautions that, while women's multiple roles and identities may be a strength in dealing with the 'messy realities of everyday activism', 'selective mobilisation' (Leckie is using Chhachhi and Pittin's (1996: 101) phrase) of aspects of women's identities may also 'stifle gender resistance' while overcommitment to several causes at the same time 'may lead to problems of insufficient time or resources'.

The interplay between ascribed and lived identities is central to the experiences of women in a wide range of different contexts across the Asia-Pacific region as highlighted in several chapters in this volume. In Rita Gallin's analysis of rural Taiwan, where Confucian ideology has been actively reinforced by the state, women become adroit at defining their identities in the interstices of conflicting perspectives: on the one hand taking into account concepts and values framed by Confucian ideals of what constitutes women's wifely and maternal roles, and on the other hand being fully cognisant of 'the more immediate realities of a woman's personal experience' (Gallin, quoting Anderson and Jack 1991: 11). Gallin argues that it is by steering a path while balancing what appear to be conflicting identities that the women of Hsin Hsing village 'engage power' to 'resist the patriarchal terrain of everyday life'. By engaging in discursive practices including 'talk' and 'gossip' as well as the public sharing and airing of disgruntlements in temple grounds to produce different 'transcripts', Gallin shows that older women contest notions of 'self' which culture has defined for them, sometimes inverting gender norms by portraying their menfolk as 'ineffectual and lazy wastrels' and themselves as 'resolute and skilled workers who engage in paid production to support the family'. As Love (1991: 96) observes, 'whereas vision [as a way of knowing] marks differences with fixed boundaries, voice establishes connections across space'. Gallin illustrates this point by showing that when women talk, they connect with one another and are 'actively resisting forms of oppression and dreaming of a better world'. While younger women appear more circumspect in using their vocality in public, they are equally adept at using more ambivalent and reconciliatory tactics in chipping away at hegemonic discourses about gender identities and the world in which they live.

In a different context, as discussed in Rebecca Elmhirst's chapter, women involved in the Indonesian government's transmigration resettlement programme in North Lampung, Sumatra, find themselves coming under the purview of state *ibuism*, a dominant discourse which represents women as 'housewives or mothers of development … firmly located in the domestic realm'. As they are denied any other identification apart from that of being 'dependants' of male breadwinners (to whom land title and assistance are given), the women find themselves excluded from title to land, as well as formal and informal sources of credit and other forms of government assistance. In fact, performing work such as 'hoe[ing] like a man' in order to feed the family subjects transmigrant Javanese women in Lampung to shame and censure for failing to conform to a feminine ideal.

Unlike the women of Hsin Hsing village who counter patriarchal discourses about a woman's place by becoming 'vocal subjects' (and hence activating the 'power' of traditional Taiwanese sayings such as 'Three women can make a market' or 'Three women produce talk [while] three men produce smoke'), Elmhirst argues that, in the context of Indonesia, female agency is less likely to be expressed through gender-specific articulations and similarly, gender identity *on its own* is unlikely to become a space for political mobilisation of women. Silence, however, does not signify an absence of struggles over gender identity.

Instead, it must be recognised that gender politics are often entwined with other projects organised around other axes, thus creating further specificities. As such, 'women' can no longer be taken as a self-evident and straightforward category and, instead, differences of interest and aims and a lack of uniformity in the foci of struggle need to be taken into consideration. Identity politics of this sort hence accepts and opens up space for 'pluralized discourses of marginalization and repression' without losing one's own perspective (Squires 1999: 135). Yuval-Davis (1998: 185) calls this 'transversal politics', a style of politics which does not assume a homogeneous point of departure or privilege a single voice, but instead respects difference while striving towards the possibility of constructing common political interests as women.

The need to recognise multiple intersecting subject positions is evident when exploring women's everyday lives for, as Elmhirst puts it, 'lived experience generally exceeds class, gender or "ethnic" categories as the meaning of these and the experiences and goals of women are variously shaped by ideologies of religion, development and nationhood in particular ways at particular times and places'. For the women of North Lampung, for example, gender interests are not articulated in 'clear and easily specified ways' but instead 'emerge within and through discourses associated with regional formations such as a reinvigorated Islam or through the idea of an emergent Southeast Asia economy, or through village-level discourses around social moralities and obligations, each expressing particular versions of femininity'. This does not mean that gender is unimportant as a mobilising centre for identities; instead, what it does suggest is that gendered subjectivities are inextricably bound up with wider cultural struggles over resources and representations.

As Elmhirst goes on to show, in the people's attempt to refuse the assimilating and unifying aims of state discourse on transmigration, the cultural politics around the issues of boundary maintenance between the 'transmigrant' Javanese and the 'indigenous' Lampungese are played out by generating opposing 'transcripts' about the behaviour of the 'other' women: while the Lampungese distance themselves from the Javanese by pointing to the 'free', 'unbridled' and therefore 'licentious' behaviour of Javanese women working in the fields, the Javanese in turn consider themselves to be 'more advanced' by pointing to the seclusion of Lampungese women and the constraints under which they live. Gender politics around issues of identity are hence inscribed in and through the politics of 'ethnic' difference in indissoluble ways.

Using women's behaviour as an ethnomarker to differentiate 'self' from 'other' takes on a more complex turn as young women (Lampungese but also increasingly Javanese transmigrants as well) leave their villages to take up work in the textile factories of Greater Jakarta, often in the face of parental disapproval. Elmhirst interprets this as a refusal on the part of the younger generation to partake in the gendered cultural politics of identity maintenance in which the transmigration area is enmeshed. Instead, they are drawing on 'an inter-regional version of "Southeast Asian factory girl" femininity' (Ong and Peletz's term) threaded through with notions of being 'fashionable', 'articulate', 'relatively worldly' and 'relatively well-educated' not only to challenge parental authority steeped in local representations of gender, but also in counterpoint to state-endorsed *ibuism* as to what constitutes appropriate femininity. In as much as the classed nature of gender politics has been pointed out (see, for example, the essays in Sen and Stivens, 1998), Elmhirst's analysis, as well as Gallin's, clarifies the importance of also taking into account how generational differences inflect identity struggles and gender politics in a region marked by rapid change from one generation to the next.

The use of 'flight' from one's natal community to 'fight' constricting gender norms and identities noted among young Indonesian women in Elmhirst's chapter is a central thread in Nobue Suzuki's account of Filipino women who sever affective ties to the family and resort to marriage or labour migration to escape gendered surveillance and sexual violence. Trapped in a masculinist regime with double standards – women's sexual purity and chastity is tightly controlled while men's overstepping of sexual boundaries through extramarital affairs and rape is often overlooked – these women decide to 'gamble' with the risks involved in unmooring themselves from 'home' to 'navigate a potentially "suicidal voyage" on the unfamiliar sea of migration'. While marriage to Japanese men or working as 'entertainers' in Japan may not always offer the emancipation that the women seek (and in fact brings with it a whole new set of gender negotiations), Suzuki shows that they find empowerment through breaking 'hegemonic constructions of the unwed daughter as virginal and dutiful' and finding 'alternative moralities' to counter patriarchal control and remake their agency in striving for 'new life chances in an otherwise unkind world'.

## From personal agency to collective activism

As seen, identity politics are flexible and fluid, often involving strategic choices about self-identifications and reworking these choices *vis-à-vis* hegemonic or competing ideologies and discourses within specific contexts. In engaging with identity politics, women's strategies range from 'silence' to giving 'voice', from subtle or symbolic (re)presentations of the body to individual action such as migration (as in the case of the young Indonesian women who leave their villages to work in factories and the Filipino women who 'gamble' with marriage to Japanese men) or simply continuing with everyday routines in the face of material restrictions or moral censure (as in the case of Singaporean Chinese women who defy cultural strictures on movement during pregnancy). Under certain circumstances, women may also resort to affirmative action with a more public face to counter state, capitalist and patriarchal dominance, demonstrating the flexibility of female power. A number of chapters in this volume address the important question of how, and under what conditions, individual self-help strategies may be transformed into collective projects, pushing and redefining the boundary between the public and the private, the political and the personal. They do not, however, provide a single answer as to how, when, where and why the personal agency of individuals coalesces to lead to collective action, but suggest that the processes involved must be contextually understood.

In the Marsinah case, Weix shows that it was the discursive reconstructions and representations thickly surrounding the violent death which allowed the transgressed body to be drawn upon in different ways by different individuals and groups. The most durable among the collective actions precipitated was the memorialisation of Marsinah's death and the elevation of her gravesite to a place of pilgrimage among young, unmarried female workers, women who are rarely considered a social force in Indonesian society and who are often unacknowledged in public. Weix argues that such concerted memorialisation has social transformative power for when the female workers 'act collectively to remember Marsinah through public pilgrimage, they draw attention to femininity, youth, and other distinctive features which define their social status outside the workplace'.

In contrast, for Haque there is no one galvanising moment or event which defines the transition into collective action; instead, she argues that women's empowerment is an embodied process, as opposed to an end-product, and examines the experience of moving from individual agency to collective action among the women who have joined the Gonoshasthaya Kendra. She shows that even as the act of joining the NGO is usually motivated by the desire to acquire new skills from training programmes on offer and not by any conscious need to advance gender interests, the coming together of Bangladeshi women from both diverse and similar backgrounds in a specific space allows them to develop connective tissues simply through the opportunities afforded by 'talking with others, exchanging ideas, learning from each other, consoling and giving each other advice'. From here, the women develop the confidence to be seen and heard, at first within the organisation itself and then out in the public places in their villages. In time, the strategies of visibility and vocality are writ large, culminating in their participation in national women's marches with clearly gendered agendas of protesting against male dominance and violence against women.

As Mohanty (1995: 69, cited in Panelli, this volume) points out, the sense of political agency must be treated as historically and locationally contingent. The process of transposing individual agency onto collective practices to further gender interests seems to follow a more discursively constructed and consciously laid out path among Women in Agriculture participants in Australia in comparison to Gonoshasthaya Kendra members. Panelli argues that, drawing on their own individual frustrations of not being readily recognised as 'farmers', participants in the movement 'have developed specific readings and sensitivities to their positions within agriculture and rural society' and 'have also come to "read and know" the dominant discourses of masculinity and market-oriented production that shape much of their experience in agriculture'. While bodily presences and presentations that make women 'visible' in the contested landscape mark the edges of gender politics in Haque's analysis, 'becoming visible' in the case of the women farmers which Panelli gives attention to is strategically promoted, not only through coming together in farm gatherings and conferences, but also in developing 'voice' through incorporating 'women's stories' into media or government events and texts, deconstructing legal and scientific discourses which impinge on agricultural knowledge and farm practices, and elaborating on the developing discourses through building social relations with one another. Discursive reflection and negotiation hence provide a range of important mobilising strategies for Women in Agriculture activists, which is further built upon through 'pragmatic affiliations' (Panelli, quoting Gilkes 1988: 68) with mainstream farmer organisations, industry boards and government departments. Panelli concludes that these strategies have been successful in 'speaking' and 'writing' women into agriculture, not just increasing their participation in their family, farm and industry settings but also expanding their power in a range of personal, industry and public spaces and channelling their influence on policy arenas formerly in the hands of the government and conservative farming bodies.

The need for mobilising strategies to span differences among women with widely different subject positions is also addressed in Leckie's chapter, albeit using different conceptual tools. Leckie notes that in post-coup Fiji 'ethnicity, nationalism and tradition are especially powerful identities intersecting with gender', and as such the resultant landscape is one littered with a heterogeneity of identities. Women in Fiji are hence not a unified activist group but find themselves sharing commonalities in the form of a 'social series', a concept developed by Iris Marion Young (1995) to describe a social collective which does not necessarily have 'shared attributes or a cohesive identity' but is drawn together as a 'loose unity' as a result of 'a shared passive relationship to a material milieu'. Leckie argues that women's agencies and activisms run the gamut from individual acts of covert protest (such as rubbing lipstick on factory clothes or deliberately assembling garments incorrectly as 'resistance' to the labour process) to embracing 'passive seriality' through membership of various organisations and trade unions, often alongside males, and further on to conscious mobilisation as a social group with public presence to carry out advocacy work to deal with gender-specific issues such as sexual violence. In the main, she argues that because of the power of ethnic, religious and traditional as well as class identities in Fiji, mobilisation in response to consciousness of common constraints *as women* often, at best, takes the form of seriality. Yet she remains hopeful that, given protracted crises and accelerated change in Fiji, women's activisms will rise beyond being a 'series of convenience' to become more active negotiations targeted at securing concrete gains for Fijian women across ethnic boundaries.

## Transnational activisms

As embodied practices, the specificities of gender politics are often well grounded in local contexts. This does not mean that 'local politics' are not also linked to, and inflected by, the multiplicity of other scales at which 'politics' may work. While most of the chapters in this volume examine how women's agencies and activisms develop within the boundaries of a specific nation-state, two of the chapters suggest that it is also necessary to trace transnational connections between activisms and across different histories and geographies.

Vera Mackie takes pains to show that this is a logical approach, in response to the fact that the seeds of oppression against which women mobilise are often 'created through the histories [which] connect people in different nations'. In her analysis of the 'spaces of difference' formed as a result of the insertion of immigrant others into the once-thought-to-be homogeneous fabric of Japanese society, Mackie reminds us that 'the relationships between immigrants and their relatively privileged hosts in Japan have been shaped by a history of imperialism and colonialism and the features of the contemporary political economy of East Asia'. The Japanese nation-state has to move from a time in the 1970s when it could assume that embodied encounters with 'difference' in the form of Southeast Asian women could be safely displaced offshore (as played out in sex tourism and other sexualised practices of 'gazing' on the rest of Asia), to having to confront the presence of these 'others' within its own boundaries. Immigrant women who enter Japan through labour and marriage migration are often marked by sexualised images, a construction which Japanese immigration policy is complicit in producing, for immigrant female workers are barred from being employed as domestics and are limited to entering the country under the legal category of 'entertainer', which is often a mask for the provision of sexualised activities from singing and dancing, waitressing and hostessing, to prostitution.

In response, Japanese activist groups such as the Asian Women's Association situate the aims of their activism within the broader context of Asia, seeking 'narratives of liberation in the histories of Third World women' in order to imagine 'a form of resistance they could not find in their own history'. Mackie shows that in dealing with 'spaces of difference' resulting from the placing of immigrant women within the nation, these activist groups have not only attempted to counter negative representations of immigrant workers and provided material assistance (such as the setting up of shelters for immigrant workers), they have also 'taken their struggles back into international circles through such forums as the United Nations Conference on Women in Beijing in 1995'. This has helped to place the living conditions of immigrant labour in Japan on the international agenda, alongside the issue of military prostitution during the Second World War and the problem of militarised sexual violence in Okinawa.

Also concerned with similar questions as to whether advocacy work among women's NGOs on migration issues can remain nation-bound, Lisa Law examines what she calls 'the emerging spaces of transnational cultural production in the realm of political activism'. Unlike Mackie, who focuses on the activist groups constituted by the host society working for immigrant others, Law locates the activist discourses and networks forged among NGOs for migrant women of different nationalities present in Hong Kong to illustrate 'the tensions between nationalism and transnationalism in the global cultural economy'. Drawing on several campaigns spearheaded by a coalition of NGOs in Hong Kong to address the protection of migrant workers' rights, Law shows how NGOs draw connections between the specificities of labour migration issues with government policies in both the Philippines (the sending country where the majority of female domestic workers in Hong Kong originate) and Hong Kong (the receiving country), as well as global discussions about human rights ventilated at the Beijing Conference. In this sense, NGOs operate as 'transforming terrains which expand the discursive field of their activities', and this in turn connects up with and potentially influences global discourses on human rights. In a globalising world where the nation is 'no longer the key arbiter of important social change' (Appadurai 1996: 4, quoted by Law), such a conception points to multiple sites of transnational activism within 'a broader social space where new alliances between migrant, feminist and workers' organisations' may be envisaged. At the same time, it has to be remembered that while these 'terrains' allow for coalition-building – for example, when domestic worker groups of five different nationalities (Filipino, Thai, Indonesian, Sri Lankan and Nepali) came together to protest against minimum wage cuts – they also decentre other nationally bound issues (such as differential wages among domestic workers of different nationalities) and may be themselves fraught with inequalities. If indeed these transnational terrains and advocacy networks give shape to a 'post-national', 'diasporic public sphere' as emergent forms of political spaces, it is likely that they will continue to reflect the tensions between national and transnational politics.

## Conclusion

As Stivens (2000: 24) has noted, 'we are not dealing with one global version of modernity, but multiple, divergent modernities within a globalizing whole. These modernities have generated their own specific, situated politics, including feminist or "womanist" politics and struggles for sexual [and other] rights.' Diverse groups of women in the Asia-Pacific, like their counterparts all over the world, generate their own particular emanicipatory politics at all levels and in all spheres. While it may be tempting to make generalizations about their resistances – as well as their resilience – in the face of existing power relations in society (for example, as a counter-offensive against the institution of patriarchy), it is clear that gender politics in the Asia-Pacific region are also highly contingent upon and contextualised by a host of other issues, most particularly questions of tradition, generation, ethnicity and nationalism. The heterogeneity of women in the Asia-Pacific is nowhere made clearer than in the difficulties of overcoming differences while forging commonalities among women through the formation of activist groups, whether locally based or transnationally stretched, in the many case studies highlighted in this volume.

Collectively, what the chapters in this volume of essays convincingly demonstrate is the need to give full weight to the specificities which inflect the particular conditions in which women of the Asia-Pacific find themselves, and at the same time grapple with the inescapable sense that despite these individual circumstances, 'gender' matters and continues to make a difference in the midst of change. What is also clear is that, given the multiplicity of modernities in which Asia-Pacific women find themselves, the complexity of women's strategies, the diversity of sites for action, and the flexibility of their alliances will continue to invigorate the Asia-Pacific scene as women carve out niches for themselves in what are still largely patriarchal worlds.

## References

Abu-Lughod, J. (1990) 'The romance of resistance: tracing transformations of power through Bedouin women', *American Ethnologist* 17, 1: 41–55.

Appadurai, A. (1996) *Modernity at Large: Cultural Dimensions of Globalization*, Minneapolis, Minnesota: University of Minnesota Press.

Callard, F.J. (1998) 'The body in theory', *Environment and Planning D: Society and Space* 16: 387–400.

Das, D.K. (ed.) (1996) *Emerging Growth Pole: The Asia-Pacific Economy*, Singapore: Prentice Hall.

de Certeau, M. (1984) *The Practice of Everyday Life*, Berkeley: University of California Press.

Dickey, S. and Adams, K.M. (2000) 'Introduction: negotiating homes, hegemonies, identities, and politics', in S. Dickey and K.M. Adams (eds) *Home and Hegemony: Domestic Service and Identity Politics in South and Southeast Asia*, Ann Arbor: University of Michigan Press.

Edwards, L. and Roces, M. (2000) (eds) *Women in Asia: Tradition, Modernity and Globalisation*, St Leonards, NSW: Allen & Unwin.

Foucault, M. (1980) 'The politics of health in the eighteenth century', in C. Gordon (ed.) *Michel Foucault: Power/Knowledge, Selected Interviews and Other Writings, 1972–1977*, Brighton: Harvester Press.

Harvey, D. (2000) *Spaces of Hope*, Edinburgh: Edinburgh University Press.

Hays-Mitchell, M. (1995) 'Voices and visions from the streets – gender interests and political-participation among women informal traders in Latin-America', *Environment and Planning D: Society and Space* 13, 4: 445–69.

Hilsdon, A.M., Macintyre, M., Mackie, V. and Stivens, M. (eds) (2000) *Human Rights and Gender Politics: Asia-Pacific Perspectives*, London: Routledge.

hooks, b. (1990) *Yearning: Race, Gender, and Cultural Politics*, Boston: South End Press.

Huang, S., Teo, P. and Yeoh, B.S.A. (2000) 'Diasporic subjects and identity negotiations: women in and from Asia', *Women's Studies International Forum* 23, 4: 391–8.

Keith, M. and Pile, S. (1993) 'Introduction: the politics of place', in M. Keith and S. Pile (eds) *Place and the Politics of Identity*, London and New York: Routledge.

Lefebvre, H. (1991) *The Production of Space*, translated by Donald Nicholson-Smith, Oxford: Basil Blackwell.

Love, N.S. (1991) 'Politics and voice(s): an empowerment/knowledge regime', *Differences: A Journal of Feminist Cultural Studies* 3, 1: 86–103.

Ram, K. and Jolly, M. (1998) *Maternities and Modernities: Colonial and Postcolonial Experiences in Asia and the Pacific*, Melbourne: Cambridge University Press.

Randall, V. (1998) 'Gender and power: women engage the state', in V. Randall and G. Waylen (eds) *Gender, Politics and the State*, London and New York: Routledge.

Robison, R. and Goodman, D.S.G. (eds) (1996) *The New Rich in Asia: Mobile Phones, McDonald's and Middle-Class Revolution*, London and New York: Routledge.

Routledge, P. (1997) 'A spatiality of resistances: theory and practice in Nepal's revolution of 1990', in S. Pile and M. Keith (eds) *Geographies of Resistance*, London: Routledge.

Scott, J.C. (1985) *Weapons of the Weak: Everyday Forms of Peasant Resistance*, New Haven: Yale University Press.

Sen, K. and Stivens, M. (1998) *Gender and Power in Affluent Asia*, London: Routledge.

Squires, J. (1999) *Gender in Political Theory*, Malden, MA: Blackwell.

Stivens, M. (2000) 'Introduction: gender politics and the reimagining of human rights in the Asia-Pacific', in A.M. Hilsdon, M. Macintyre, V. Mackie and M. Stivens (eds) *Human Rights and Gender Politics: Asia-Pacific Perspectives*, London: Routledge.

Thrift, N. (1997) 'The still point: resistance, expressive embodiment and dance', in S. Pile and M. Keith (eds) *Geographies of Resistance*, London: Routledge.

Vidler, A. (1978) 'The scenes of the street: transformation in ideal and reality, 1750–1871', in S. Anderson (ed.) *On Streets*, Cambridge, MA: MIT Press.

Waylen, G. (1998) 'Gender, feminism and the state: an overview', in V. Randall and G. Waylen (eds) *Gender, Politics and the State*, London and New York: Routledge.

Wille, C. and Passl, B. (eds) (2001) *Female Labour Migration in South-East Asia*, Bangkok: Asian Research Centre for Migration.

Yuval-Davis, N. (1994) 'Identity politics and women's ethnicity', in V.M. Moghadam (ed.) *Identity Politics and Women: Cultural Reassertions and Feminisms in International Perspective*, Boulder, Colorado: Westview Press.

——(1998) 'Beyond differences: women, empowerment and coalition politics', in N. Charles and H. Hintjens (eds) *Gender, Ethnicity and Political Ideologies*, London: Routledge.

# 2 Nine months

## Women's agency and the pregnant body in Singapore

*Lily Phua and Brenda S.A. Yeoh*

## Introduction

Feminist scholars have recently argued for 'a new epistemological viewpoint based on the idea of knowledge [and hence power from a Foucauldian perspective] as embodied, engendered and embedded in the material context of place and space' (Duncan 1996: 1). In this context, and given geographers' age-old preoccupation with space, it should come as no surprise that feminist geographers are beginning to interrogate the body as a terrain for oppression and resistance at the most minute scale. By exploring 'the ways in which bodies themselves are imagined as spaces, and the spaces they are imagined as inhabiting ... in relation to a range of subjective, emotional and psychic processes' (Gregson *et al.* 1997: 196), the body assumes a critical role in understanding the politics (including gender politics) of everyday life. In a parallel vein, critical medical geographers are also beginning to recognise that 'attention to the embodied subject and a concern with theorising identity and difference provides a language and set of ideas that further the artic-ulation of the complex links between place, space, power relations and health' (Dyck and Kearns 1995: 140).

Feminist geographers have primarily worked towards showing that *the* body is an illusory masculinist composition; *the* body does not exist. What exists are *bodies* (Rose 1993; emphasis added). Furthermore, these bodies are sexed (Johnson 1989). By examining the notion of the body in its representational and material forms, femi-nist geographers are gradually recognising the body as 'a vital element in the constitution of masculine and feminine identity' (Johnson 1990: 18), and as 'both constituent and conveyor of social, political, and psychological meanings' (Moss and Dyck 1996: 747). This chapter focuses on the politics of the pregnant body, a bodily form which only women assume. As Longhurst (1998: 21) has argued,

> studying pregnant women offers the possibility for disrupting masculinism in geography ... in that pregnant women effectively illustrate the notion of Other being Self and Same. Pregnant women undergo a bodily process that transgresses the boundary between inside and outside, self and other, subject and object. This serves to problematise the framework of binary opposition through which the authority of key concepts is established in geography.

The primary aim of this chapter is to elucidate the vibrancy of human agency and embodied reflexivity among pregnant Chinese[1] women in Singapore. Like all other societies characterised by forms of unequal gendered relations, multiple structural forces – expressed as dominant state, medical and cultural discourses and executed through specific strategies of control – stake a range of claims on pregnant bodies in Singapore. However, as active, living and reflexive bodies, pregnant women confronting the weight of hegemonic forces are capable of engaging in a panoply of strategies – sense-making, negotiation, resistance and sometimes acceptance and reinforcement – through their spatial practices as traced in everyday life. The material for this chapter draws on a two-year (1997–8) research project exploring local women's experiences of pregnancy. The findings are mainly based on an analysis of 338 questionnaires resulting from a survey[2] and twenty-seven in-depth interviews[3] with Chinese women who were either pregnant or had given birth to their youngest child within the twelve months prior to the interview.

## Power and the pregnant body

In discussing women's embodied agency, the notion of power as propounded by Michel Foucault and expanded by other social theorists is particularly instructive. From a Foucauldian perspective, in order to understand the play of power relations in modern society, attention to the body (in the plural sense) is crucial as power operates upon, through and by the body. The body is a terrain for discipline and resistance, a site upon which structural forces and practices impinge, and an operant of power as it reflexively undertakes the performance of individual practices (Foucault 1980a; Turner 1996). Put simply, the body is the site, medium and operant of power. At the same time, recognising that individuals not only *have* bodies but also *are* bodies (Turner 1984), power and agency must be viewed in relational terms. Power is a relationship between active persons and is only sustained in so far as each is 'thoroughly recognised and maintained to the very end as a person who acts' (Foucault 1982: 220). What this implies is that power is only possible in the presence of human agency, and through embodied actions. As Foucault (1980a: 95) puts it, 'where there is power, there is resistance'.

Just as power is exercised over the pregnant body through strategies of control – 'dispositions, manoeuvres, tactics, techniques, functionings; that one should decipher in it a network of relations, constantly in tension, in activity' (Foucault 1979: 26) – resistance can also be effected by the pregnant body. Resistances to power

> are all the more real and effective because they are formed right at the point where relations of power are exercised; resistance to power does not have to come from elsewhere to be real, nor is it inexorably frustrated through being the compatriot of power. It exists all the more by being in the same place as power.
>
> (Foucault 1980b: 142)

As Annandale and Clark (1996: 31) put it,

> The dominant discourse ... must itself create the conditions, or discursive spaces, for a reverse or alternative form. Indeed, the very existence of the dominant form depends on points of resistance to act as a target and support. So power is a resource for action and it is possible (or, perhaps, even necessary) to recognise areas ... as contested sites in both contemporary and historical form ... to move away from a passive conceptualisation of women controlled by obstetrics (while still recognising the institutional power of dominant discourse), and to presume the co-presence of a contested voice.

It follows from here that just as the exercise of power permits domination and control, it also defines 'innumerable points of confrontation, focuses of instability, each of which has its own risks of conflicts, of struggles, and of at least temporary inversion of power relations' (Foucault 1979: 27). In fact, the existence of power 'depends on a multiplicity of points of resistance: these play the role of adversary, target, support, or handle in power relations' (Foucault 1980a: 95). This view has been similarly advocated by Giddens (1981: 4) who argues that 'power and freedom in human society are not opposites; on the contrary, power is rooted in the very nature of human agency, and thus in the freedom to act otherwise'. Hence, even 'the most dependent, weak and the most oppressed ... have the ability to carve out spheres of autonomy of their own' (Giddens 1987: 11).

Foucault also acknowledges that, like power, resistance is diverse and can be integrated in the form of strategies. In the words of Crow (1989: 4), 'strategies are not exclusively the preserve of dominant groups. Dominated groups, too, may devise strategies, perhaps in response to those of dominant groups which impinge on them.' Crow (1989: 4) has gone on to suggest that 'some of the most sophisticated strategies are those developed in response to the strategies of others'.

From this perspective, pregnant women as 'living, speaking and laboring' (Foucault 1985: 7) bodies are capable of resisting attempts of domination and control through devising and executing strategies of resistance. The pregnant body is thus not a product placed exclusively in the hands of dominating individuals, social groups or systems in society, but also one which women deploy as a resource and weapon to resist and negotiate power relations. At the same time, not only do pregnant women have pregnant bodies, they also *are* pregnant bodies. As Longhurst (1997: 10) puts it,

> Pregnant bodies, like all bodies, are an interface between politics and nature, and between mind and matter. They are 'real', while at the same time, they are socially constructed. 'Real', material pregnant bodies don't exist outside of the political, economic, cultural and social realms.

Strategies of resistance denote the actions individual women take 'that attempt to challenge, change or retain particular circumstances relating to societal relations, processes, and/or institutions' (Routledge 1997: 69). These embodied strategies which may effect contestation, modification or rejection can be attitudinal or behavioural, discursive or practical in character, and are more often than not executed in the course of everyday routines and activities in time and space, through what Lefebrve (1991) has termed 'spatial practices'. Spanning a broad spectrum of styles, these strategies of resistance may assume more active forms, such as outright rejection or direct conflicts, or more passive styles, such as concealment, evasion and feigned compliance. In the context of Singapore, while overt, outright, collective protests against structural domination over their pregnant bodies are rare, there are many forms of unspectacular, subtle and covert embodied strategies engaged by pregnant women in their everyday lives to counter, negotiate and inflect external societal control over their bodies. These everyday forms of resistance (Scott 1985), while seldom capable of bringing about a complete overhaul of existing webs of power and domination, may be 'fundamentally more radical and effective responses to the deployment of power' (Martin 1988: 10) in the course of daily living. Even if they may not be able to effect immediate changes, they may induce various institutions of social control to modify their discourse over time.

## Procreation and pregnancy in Singapore

Pregnancy in modern Singapore society is very much a bodily condition governed by multiple structural forces. Certain patriarchal notions of the reproductive and/or pregnant body find expression in intersecting state, medical and cultural discourses. A closer look at each of these dominant discourses in Singapore provides one means of unveiling the patriarchal face of the state. As will be argued, these discourses constantly frame the reproductive and/or pregnant body as one of a number of different bodies requiring constant surveillance and regulation for the advancement of the family, society and, ultimately, the nation (for a discussion of the regulation of other bodies, see Bungar 1991; Yeoh 1995/6; Yeoh 1997). Once the pregnant body becomes defined and normalised as such, a whole range of strategies of control (discursive and material) are legitimised and effected. By defining the pregnant body as problematic and by offering 'solutions', the exercise of power is made possible as the establishment and execution of a whole range of strategies of control in everyday life is legitimised.

In Singapore's post-colonial history, public statements have often emphasised Singapore's fragility as a nation with limited resources. Accompanying this rhetoric of national survival are narratives of 'national crisis', issues perceived to have the potential to threaten the country's continued existence. It is interesting to note that 'population' and 'reproduction' often occupy a central position in these narratives. Often starting out from the standpoint that population is the only resource Singapore can draw upon to ensure its economic vitality, state narratives seldom fail to propose 'procreation' and 'women' as 'problems' that deserve the state's 'pastoral' gaze.[4] As noted by a number of scholars (Purushotam 1992; Heng and Devan 1995; Kuah 1997), state narratives have specifically targeted women and their non-compliant bodies as the crux of the country's population woes. In recent years, whenever this so-called 'national problem' is discussed in parliament or in newspaper forum pages, women – whether single women whose 'orientation' needs changing 'from career to family' or wives who are said to 'call the shots on the number of babies' – are seen to be 'the root of the problem'. Women's groups and some women members of parliament have counter-argued that it was 'the men who needed their mindsets changed' to equalise responsibilities over parenting and the house-hold (*Straits Times* 2000a, 2000b).

Despite several pendulum swings in the course of the last thirty-five years, state discourses have continually defined the pregnant body as problematic – first as over-productive, then as selectively over- and under-productive, and more recently as under-productive (as reflected in the three major phases in the country's population policies as well as by the changing positions assumed by state officials and agencies over the years, from an anti-natalist stance to a eugenic one, and later to a pro-natalist stance; see Teo and Yeoh 1999 for further elaboration). By repeatedly framing the pregnant body as 'a problem' that has repercussions for the nation, state intervention is at once ratified and its applica-tion of a wide array of instruments to harness state surveillance of the pregnant body is readily legitimised. Hence, through the years, the attempt to gain ideo-logical hegemony has been pressed through a variety of strategies, including massive 'family life' campaigns aimed at motivating reproductive bodies to behave in ways deemed 'appropriate' by the state (see Figure 2.1); the allocation of state resources (such as subsidised hospitalisation fees and child-care subsi-dies); and the institutionalisation of disciplinary practices through state policies (such as education and housing policies), legislation (such as the Abortion Act and Sterilisation Act) and various state establishments (such as the Singapore Family Planning and Population Board, Social Development Unit and the Family Life Education Coordinating Unit, each with their own programmes and activities to facilitate surveillance of women's reproductive bodies).

So too has medicine, with its associated systems of knowledge and routine practices, staked claims on women's pregnant bodies. In Singapore, under-standing of the pregnant body is steeped in the discourse of 'medicalisation'.[5] A battery of ante-natal medical examinations at clinical institutions based on western biomedicine and with occasional recourse to Chinese biomedical prac-tices are a very much taken-for-granted aspect of individuals' experiences of pregnancy. While there exist many differences in the conception of pregnancy between western biomedicine and traditional Chinese medicine, both forms of medical knowledge share a common biological–medical language propounding the 'monitoring of health/normality', 'detection of abnormality', 'medical inter-vention' and 'identification of risk factors' with the expressed objective of delivering 'healthy' babies. Generally, dominant discourses of pregnancy in both

## Addressed to the single woman

*Are you giving men the wrong idea?*

Are you giving men the wrong idea? It's wonderful to have a career and financial independence.

But is your self-sufficiency giving men a hard time? They say that you expect a lot from them and have become intimidating and unapproachable.

Surely that can't be true. You really are a warm and friendly girl, and look forward to a home of your own and a family.

Perhaps it's time to give the guys a break. By being more relaxed and approachable. Friendlier and sociable. That way, they'll get to know you – which is how relationships begin.

After all, you don't want to give men the wrong idea.

## Addressed to the single man

*Do you keep up with the times?*

Do you keep up with the times? If you're going for success in life, you have to keep up with the times, right?

But when it comes to your relationships with girls, does the same apply? Or are you in the old mode and chauvinistic in preferring girls who aren't your equals, who will be awed by you and be at your beck and call?

If it is true, you aren't keeping up with the times. For a man needs a partner, someone to give help and encouragement, someone you can be proud of (just as she's proud of you). That someone is most likely to be your social and intellectual equal. So chat up the girls. Make friends with them. That way, you'll get a real partner in life.

*Figure 2.1* Texts in a Singapore family life poster advertisement used in 1988

knowledge forms place pregnancy as a biological and medical event. These biological–medical discourses not only present pregnancy as a form of bodily 'weakness', if not 'illness', requiring medical regulation and intervention, they also justify the penetration and manipulation of women's inner bodily spaces with medical technologies, albeit to different extents. In addition, the rise of 'surveillance medicine' (Armstrong 1995) resting on the notion of 'risk factors' furthers the social control of women's bodies through positioning them in 'an extra-corporal space – often represented by the notion of lifestyle' (Armstrong 1995: 401). As biological–medical discourses revolve around the possibility of risk factors affecting the health of pregnant bodies and eventual delivery of healthy babies, pregnant women's everyday spatial practices – daily dietary and hygiene habits, smoking and drinking activities, and physical exercise routines – all come under the purview of the medical gaze.

In a multicultural society like Singapore, the pregnant body is also inscribed in culturally specific ways. Among the Chinese, two main sets of concepts serve to inform cultural understandings of the pregnant body. First, in a predominantly patrilineal Chinese community, married pregnant bodies are typically accorded high social status. Pregnant Chinese women usually earn social esteem as procreators, as their bodies carry the possibility of extending the paternal lineage. This is especially so if the foetuses they carry within themselves are male. This strand of discourse is marked by a series of cultural attitudes, expectations and behaviours which serve to separate women's 'non-pregnant' everyday lives from their 'pregnated' everyday lives. Seen as *xishi* (a Chinese term connoting an auspicious and joyous event), pregnancy is usually seen as a culturally desirable, fortunate and welcomed event, and pregnant women are usually accorded special treatment and privileges (also see Lim 1994).

On the one hand, the elevated treatment pregnant women expect and receive within the private spaces of the family as well as in public spaces is possibly indicative of temporary inflections of gender relations in a society generally guided by strongly ingrained patriarchal assumptions (Purushotam 1992; Phua and Yeoh 1998). On the other hand, however, women's role in serving as the embodiment of their husbands' patrilineage renders these bodies constantly under surveillance, given concerns not so much for the women themselves but for the foetuses residing within the wombs of these bodies. A second set of cultural concepts relating to pregnancy and pregnant bodies revolves around traditional Chinese understandings of health and illness which see the pregnant body as weak, not only physically but emotionally as well (Furth 1987).[6] This again serves to legitimise a range of surveillance strategies over pregnant women's everyday lives. By conceiving of pregnant bodies as intensely precious to the continuation of one's patrilineage, and at the same time as being potentially prone to the invasion of imbalance, disorder and disease, the notion of women's pregnant bodies as requiring delicate care and thorough scrutiny to guard against possible threats and dangers is continuously reinforced.

To explicate the web of power relations spun around the pregnant body in Singapore, we have briefly highlighted in turn state, medical and cultural discourses. These, however, should not be taken to imply that any single set of discourses constitutes a uniform conceptual framework 'embrac[ing] particular combinations of narratives, concepts, ideologies and signifying practices, each relevant to a particular realm of social action' (Giddens 1986: 8). Instead, these are interwoven discourses which comprise 'multiplicities of discursive elements that can come into play in various strategies' (Foucault 1980a: 100), often in overlapping rather than diametrically opposite ways.

## Body politics: an empirical investigation of pregnant bodies

As argued, in Singapore a number of dominant discourses converge to erect a regulatory regime which considerably limits and fractures the spaces within which pregnant women may manoeuvre without being labelled 'deviant'. Nevertheless, beneath the surface of apparently unruffled lives and seemingly uncontested regulatory practices, a diversity of 'low profile' resistances do occur in the realm of the everyday (Scott 1990: 198). As Hannah points out (1997: 349), in contrast to the extreme case of confined prisoners whose life paths are 'completely visible to the authoritative subject', the spatial–temporal movements of 'free' citizens such as pregnant women are, by virtue of their higher degree of spatial freedom and anonymity, much less visible. This allows for greater possibility to escape complete surveillance and discipline. While Hannah (1997: 352) has maintained that for 'free' citizens, 'many activities are ... regulated as much by the threat of observation as by actual surveillance', it is important to note that individual and collective subjects may also draw on everyday practical knowledge in devising counter-strategies of their own.

Through spatial practices and micro-level strategies which surge and subside in time and space, assuming different styles at various sites and spanning a spectrum from 'active' to 'passive' means of resistance, embodied individuals are capable of skilful strategies to negotiate and negate the exercise of power attempted over their bodies and daily embodied experiences and practices. As will be shown, these strategies of resistance are largely incapable of revolutionising the existing social order or introducing large-scale structural changes in society. However, it is maintained here that these persistent and multitudinous strategies of resistance as effected over a long period of time possess the potential to compel the various institutions of social control to constantly revise the discourses they embrace, as well as their own policies and strategies. At the same time, these seemingly insubstantial strategies of resistance may well produce instantaneous gains for pregnant women at the micro-level in everyday life.

The following sections go on to sieve out the multiple strategies of resistance embodied individuals undertake to negotiate diverse state, medical and cultural discourses around the notion of the reproductive and/or pregnant body. Embedded within the everyday spatial practices of pregnant women, these strategies provide a counter to representations of women as victimised, passive objects of monolithic systems of oppression.

### Contesting state discourses on reproductive bodies

Resistance to state discourses on the reproductive and/or pregnant body has been particularly salient in the last decade, as the population as a whole becomes generally better educated and financially better off. The pro-natalist New Population Policy introduced in 1987 in an attempt to reverse persistent below-replacement birth rates and ameliorate concerns in a labour-short economy has met with far more critical response compared to the pliancy characteristic of the response towards the anti-natalist Old Population Policy (Teo and Yeoh 1999). The failure to reverse 'plummeting birth rates' (the Total Fertility Rate dropped from 2.8 in 1970 to 1.48 in 1999) is considered a serious and intransigent problem amounting to 'a nation … committing collective suicide' (*Straits Times* 2000a). It is now clearly accepted that 'the slew of pro-baby incentives' had been 'sterile', and the 'procreation problem' remained 'acute' (*Straits Times* 2000b).

The promotion of a pro-natalist approach to sustain the growth of the economy has not had its desired effect, primarily because the common belief among Singapore women is that 'decisions about family size are personal and should not be a matter for government intervention' (Graham 1995: 229). While the policy has not provoked collective public disapprobation (and in fact seemed to have some level of superficial concurrence in public discourse), it has clearly stirred up social and moral indignation among individual women whose tendency is to articulate their displeasure in the private sphere. As has been argued,

> [w]hile autonomy in fertility decision-making may actually be more illusory than respondents would like to admit, the insistence on personal degrees of freedom suggests that ideological concurrence at the public level masks some degree of ideological resistance at the personal level.
>
> (Teo and Yeoh 1999: 93)

In the study, one of the more active resistant utterances came from Celine, an administrative officer in her late twenties. With a 2-year-old son and another child due within a few days of the interview, Celine was adamant that she would be 'stopping at two whatever the policy' and had 'plans' to prevent herself getting pregnant again because she considered the cost of raising more than two children unaffordable. She also maintained that the state had no right to intervene in what she considered the 'private affairs' of individuals:

> they have no right [to interfere] … I feel that they have no right to ask us to 'stop at two' [referring to the slogan characteristic of the anti-natalist era] or to give birth to more [referring to the current 'have three or more if you can afford it' policy]. It's up to us whether we can or cannot afford. And whether we can or cannot give birth. Some families … they may want more children, so if you ask them to stop … I don't think so.

In counterpoint to Celine's direct disavowal of state discourses, other more indirect, inflected forms of resistance also persist, suggesting that there are cracks and crevices in the state's ideological hegemonic control over women's reproductive bodies. Actively rejecting state discourses of reproduction or outwardly challenging the various institutionalised techniques of power have not been popular styles of resistance. Rather, the more common strategy is to combine an avowal of generalised support for state policies in the public realm (as detected through surveys and the like, see Teo and Yeoh 1999) while personally disregarding the state's call for its citizens to participate in solving the problem of low fertility rates (often by referencing other constraining factors).

For example, when asked for her opinion of state policies relating to reproduction and fertility, Nora (full-time homemaker in her early thirties) seems to
have little hesitation in concurring with the need for state intervention:

> frankly speaking, they [the government] have their own reasons. Just think
> about the population in China. It's really large. If the government does not
> control, then it will become a big problem. And a family unit will in one way
> or another affect the country's condition. So if the government should
> impose any policies, they have their own reasons. ... so we should not feel
> that they are interfering or what [*sic*].

However, in the same account, non-compliance to the state discourses is also
evident when it comes to personal reproductive behaviour:

> because I am not working ... and I feel that it is too taxing especially when
> you talk about their education. It is pressurising. ... You have to see how fast
> the child is able to learn ... If she is slower at picking things up, then you
> have to spend more time and effort on her. Then, there is another problem
> ... if they are not able to cope and be at par with their peers ... they may
> lose interest in studying. ...[Also] let's say if you have the ability, then you
> may have three ... or even four kids. But now, because of our living stan-
> dards, it doesn't permit us to have that many.

This strategy of playing off espousal of the 'public good' *vis-à-vis* private non-
compliance allows women to evade the stigma of being labelled 'inconsiderate
citizens' or 'deviant bodies' and at the same time preserve a sense of autonomy
over their own reproductive bodies.

By distancing themselves and their everyday reproductive behaviour from the
broader narratives of 'national crisis' (Heng and Devan 1995), women have been
able to engineer a strategic form of discursive exclusion of their own bodies from
the hegemonic grasp of the state. This form of resistance is also often aided by
their hinging their decisions on practical economic factors (such as high standards of living, the need to plan for the long-term educational needs of their
children, the lack of familial assistance in childcare and household duties, and
the need to engage in paid employment outside the family) as non-negotiable
parameters, so as to discharge their 'national obligation' to have more children.
In that the state makes multiple demands on women's bodies as both reproductive bodies to produce 'the creators and protectors of the next generation'
(Senior Minister Lee Kuan Yew, quoted in Doran 1996: 157) as well as labouring
bodies to turn the wheels of the (labour-short) economy, women are also able to
capitalise on these contradictory claims to assert their own personal freedoms.
Often, interviewees draw attention to what they perceive as conflicting state
discourses as a means to explain women's non-compliance. Elaine (an administrative officer in her early thirties), for example, argues that state policies are
'contradictory':

the government is trying to encourage women to give birth. However I feel that there's not much being done to support pregnant women, mothers … [especially] a young mother … [We are encouraged to] go back to work … [so that] the productivity level [will] be kept [up], you must be 'number one' and those sort of things. So you [i.e. the nation] need[s] the women labour force, so [we] 'work, work, work' … But on the other hand, the Ministry of Home Affairs or Community Development … they will say, 'Stay at home and take care of your own kids.' So there is a contradiction. But then there is not much done to help support this [i.e. staying home].

Amy (a homemaker in her early thirties) points to other contradictions:

In the past when the people wanted to have more children, they were fined. It's so funny … to me, they are a bit of a *shen jing bing* [Mandarin for 'crazy']. Last time you are supposed to stop at two. Now you are asked to try your best to give birth to more children. But now most of the Chinese all stop at two. All my friends stop at two. All of them are saying that studying [referring to children's education] is so difficult.

In so far as the state sees child-bearing and child-rearing as women's work (Yeoh and Huang 1995) and women's non-compliant bodies are seen to be 'imperiling the country's future by wilfully distorting patterns of biological reproduction' (Heng and Devan 1995: 197), women are also capable of exploiting the anxieties of a state patriarchy caught 'between reproducing power and the power to reproduce' (Heng and Devan 1995: 200) and positioning their own subjectivities in the interstices between policies. In their own ways, women subject state discourses to constant discursive negotiations, often explaining and legitimising their own non-compliance by pointing to the state's inherent 'contradictions' and interpreting them as failure to uphold its end of the 'patriarchal bargain' (Kibria 1990).

### Negotiating the medical gaze

Foucault (1980b: 167, 170) has argued that the question of power is impoverished if it is posed solely in terms of the state and that instead of locating 'the centre of initiative, organisation and control for [the] politics [of health] only in the apparatuses of the State', power should be viewed as running through diverse 'capillaries' of society where '[d]ifferent power apparatuses are called upon to take charge of "bodies"'. Medicine, with its associated systems of knowledge and routine practices, constitutes one of these power apparatuses which function to exercise surveillance over women's pregnant bodies. Indeed, medical interventions such as ante-natal examinations at clinical institutions based on western biomedicine are a very much taken-for-granted aspect of women's experiences of pregnancy in Singapore. It should also be noted that medical discourses surrounding the pregnant body are not entirely divorced from those of the state; instead, state discourses often make acceptable medical intervention among a 'population' constituted by women's bodies, which in turn renders these bodies more vulnerable to the state's demographic control.

While women's engagement with the state has mainly focused on inflecting its discourses on the reproductive body, their negotiations over the power of the medical gaze are situated more specifically in the stream of lived experiences. In the course of everyday life, women engage in struggles to thwart absolute medical control over their bodies and to redefine dominant medical understandings of the pregnant body, even as they attempt to make sense of their own bodies. This process of negotiation occurs at different sites interspersed throughout the everyday spatial practices that they trace. One such site is the ante-natal clinic.

While the ante-natal medical examination provides an important site for the exercise of power over pregnant bodies, room is also created for 'the effective resistances of the people' (Foucault 1980b: 162). At this particular focused site where the pregnant body constitutes the explicit and primary focus in the clinical encounter between women on the one hand and medical professionals on the other, struggles over the control of the pregnant body may involve ongoing, dynamic negotiation of competing understandings of the pregnant body.

Elaine recalls instances at the ante-natal clinic in which she attempted to come to grips with dominant patterns of medical intervention. When medical checks and tests were conducted, no explanation was given to Elaine as to why these were done, nor was permission asked for. The nurses who carried out these procedures (such as the weight check and a battery of blood and urine tests) usually just told her to 'do this and do that', but were unforthcoming when it came to explanation. Initially, Elaine tried to overcome this sense of uncertainty about what was being done to her body by first actively asking questions:

> I did ask them [the nurses] to give me some explanation which ... well, you know, they were mumbling ... and I couldn't really hear. So I didn't ask them any more after that.

Not having been able to draw out any satisfactory answers from the nurses, she turned her queries to other medical professionals such as the gynaecologist and obstetrician, to insist on getting some answers to her questions. Elaine also sought to actively bring to bear on her encounters in the clinic other knowledge gleaned from conscientiously reading up on a wide range of 'expert' books on pregnancy (produced by local and foreign authors) as well as consulting and comparing clinical experiences with friends who were also pregnant or had been pregnant. These reflexive steps, taken not so much to challenge but check on the integrity of the medical regime as played out during the clinical encounter, illustrate individual autonomy on the part of Elaine and her refusal to submit her own body to medical domination without questioning and evaluation. Elaine made clear her active agency in managing the clinical encounter:

I told him [the obstetrician] what *I* wanted. I told him I wanted a natural birth and I didn't want any painkillers as far as possible. … also no epidural. I told him I didn't want it.

Resisting strategies are particularly intense when there is a slippage between what medical power prescribes and women's own 'feeling' about their own bodies. In her account of her pregnancy, Wei Yuet (a clerical officer in her early thirties expecting her first child) describes how she first went to a particular gynaecologist on the recommendation of her cousin-in-law. But after a while, Wei Yuet felt that

she [the gynaecologist] was not very good … I found that she didn't give you a sense of security. Sometimes the doctor … is very important. You need to feel comfortable with them. That is what I *feel* … Because that time I went and she just said, 'You are all right. There is no problem.' But I am *not* all right! I can tell you that I am not all right. And every time I go there, she will give me a jab to stabilise my baby, and she charges me very high. And then I said, 'Okay. You can charge me that but you also need to give me some reassurance.'

As the particular medical professional did not satisfy her expectations, Wei Yuet eventually decided to stop attending the specific clinic. She checked with a number of friends for advice, and ultimately settled on an alternative ante-natal clinic in the neighbourhood where the medical advice, in her view, accorded better with her own 'instincts' about her own body.

Apart from the ante-natal clinic, the power of the medical gaze is also continually being wrestled with at a number of different sites interspersed in the course of everyday life. Many of these occur when women encounter significant others, often other women. Shu Min (a manager in her early thirties), for example, makes it a point to consult

friends who have just gone through pregnancy … I talk to them, like 'Do you feel like this?' 'Did your gynae say this, do this?' Compare notes lah [*sic*], and see if there is any way to help yourself.

A sense of 'helping oneself' may also emerge incidentally in the context of everyday conversations about women's embodied experiences. Telephone calls, for example, often serve as a means for women to express and negotiate their agency. While the primary aim of these calls may not be necessarily related to experiences of pregnancy *per se*, the context for resistance against medical power may still emerge in the course of chit-chatting or discussing other matters. Swee Hong (full-time homemaker in her late twenties and two months into her third pregnancy), for example, kept in frequent phone contact with her mother and four of her five sisters (they all live in different housing estates scattered throughout Singapore), primarily to keep in touch and maintain familial ties ('We always call one another. Just to ask what they are doing. Tell them what's happening ... We are all housewives. Too free, you know'). At the same time, however, these conversations also become grist to the mill as Swee Hong contemplates ways to manage her pregnancy, providing a social context which makes alternative strategies possible. For example, Swee Hong knew that the western-trained gynaecologist she goes to for regular consultations and check-ups had advised her against consuming traditional Chinese herbs. However, after casual conversations on the telephone with her mother and sisters revolving around how best to improve well-being, she was encouraged to buy and prepare different herbal brews to enhance her health:

> it's just that I hear my sisters and my mother say [how good particular herbs are for pregnancy] ... so I go and buy. They also teach me how to cook these [herbs] ... They give me good advice based on their own experience.

Swee Hong is not so much choosing between different medical systems to manage her body during pregnancy but selectively combining what she sees as medical advice which best suits her needs by pragmatically drawing on different sources of medical discourse. From her point of view, this is not construed as a rejection of western medical advice but a means of reflexive self-help and agency.

Negotiations over the salience of the medical gaze may take place in a wide range of unfocused, and unexpected, sites. Yvonne (a sales assistant in her early thirties) relates an encounter during one of her visits to a particular hairstyling saloon. As she was engaging in casual conversation with the shampoo girl, the topic of pregnancy and ante-natal medical examination surfaced when the shampoo girl noted that Yvonne was pregnant. According to Yvonne,

> this salon girl came and asked me ... 'When you go for check-up, do you need to check your breast?' I said, 'For what?' ... I said, 'Check breast for what?' She said, 'This gynae checks my girlfriend's breasts every time. And it's so embarrassing. He always makes her strip from the top to the bottom, even underwear, you know.' ... Then I said, 'What did he check?' 'Oh, he checked for breast cancer,' she said ... That's very strange ... that's why I told her to report but she said that the next time she wouldn't go back. ... because by the time she asked me ... well, she's so stupid to say out only in her late pregnancy ... She kept it to herself until late pregnancy, then she told me. By then she was just two more visits before delivering. She said she dare not go back to that gynae after delivery.

At sites such as this shampooing session, understandings of what constitute the parameters of ante-natal medical examinations come under scrutiny. By exchanging notes on one another's experiences, women come to question the extent to which medical intervention legitimises procedures which violate women's bodies and impinge upon women's embodied sense of propriety. As Yvonne concludes,

> these are just stupid reasons ... breast cancer. What has that got to do with pregnancy? And she [shampoo girl's friend] even let him [the gynaecologist] check again and again! He even commented on things like, 'You know, oh ... you have got very beautiful nipples. You should do breastfeeding.' Hey ... hello! Breastfeeding has got nothing to do with beautiful nipples, you know!

As a result of the encounter in the salon, Yvonne expressed a strong sense of suspicion towards male gynaecologists, insisting that

> I will make sure ... especially if it's a male gynae ... the more you have to make sure that they have a good reputation ... because some are molesters. It's true.

### Resisting cultural discourses

Apart from being the subject of state and medical interventions, the pregnant body is also culturally inscribed. Women often relate to cultural discourses surrounding pregnancy in ambivalent ways, at times buying into so-called traditional Chinese discourses and at other times evading or rejecting their grasp.

While most of the Chinese women interviewed agreed with the traditional position that pregnancy *should be* a joyous and desirable event, some feel that their own personal embodied experiences of the state of pregnancy proved otherwise. Shu Min, for example, disliked being pregnant

> because I don't go out for social visits ... I don't feel good about it. That's why I don't like being pregnant. ... It's like a lot of things ... I have problems with. But I have no choice, you see ... [I have to make] sacrifices ... no freedom lah [*sic*]. Like I said just now, a lot of things I cannot do although I wanted to ... you know, it's like not being yourself. Not being myself, and it's not a choice of mine.

She asserts that after this pregnancy (her first), she would actively take steps to ensure that she would never be placed in a position when she could not claim her body as her own again:

> definitely ... after this I am going to 'tie' (referring to the medical procedure of clipping the fallopian tubes to prevent future pregnancies). I am not going to take the risk. It's quite a traumatic experience to go through.

While some women enjoyed the pregnant state, others declared that they 'hated the nine months', 'couldn't wait to get it over and go back to what I was', 'felt that I looked awful', and generally disliked what they saw as a loss of control over their own bodies. As Chai Yun (a clerk in her early thirties) puts it,

> It's not that I don't like it [referring to the foetus she was carrying] ... but it's like being trapped for nine months!

The idea that pregnancy constitutes being 'trapped' in a body which is somehow not under one's control is compounded by other strands of Chinese cultural discourses which position the pregnant body as physically and emotionally weak. While most of the women interviewed took this point of view for granted, some took active steps to challenge such a perspective by drawing on alternative forms of knowledge about pregnancy. For example, Felice (administrative officer in her late twenties) was determined to treat her pregnant body as anything but weak:

> I read up books ... the more western ones. I think the viewpoint is different, the mindset is larger, whereas in Chinese tradition you are treated almost like a sick puppy ... you should stay at home and do nothing. I think the western view is more encouraging ... We *can* do things!

In the same spirit of insisting that the pregnant body is not necessarily weak or sick, other women also tried to counter cultural strictures on dietary issues to carve out a larger sphere of autonomy for themselves in everyday life. It should be noted that the culturally dictated taboo of avoiding foods such as crabs, mutton, chilli and pineapples[7] has widespread acceptance among Chinese women in Singapore (only 19.5 per cent of the women surveyed disagreed with this particular tradition), even if such acceptance is rooted not so much in actual belief in the efficacy of such taboos (they are but *bang dang* (superstition) and 'old wives' tales' according to several interviewees) but a desire not to do things differently from what is demanded of them by 'tradition'. (As Swee Hong puts it, 'Whether it's the truth or not, we won't know. But it's better to listen to follow what they say rather than to regret later on.') There is, however, a range of resisting strategies which take the form of outright rejection to more ambivalent tactics. The expression of resistance to the practice of avoiding certain foods during pregnancy by Chew Mee (a lawyer in her early thirties) represents those who have decided not to subscribe to the food taboos:

> things to avoid eating ... you know, stuff like mutton ... certain types of bananas ... there are just so many! I think ... the stuff like mutton and all ... there is no scientific evidence, I just didn't bother. I still ate what I liked. ... The other thing is that in traditional Chinese families, they [expect pregnant women to] take a lot of herbs, all kinds of Chinese herbs for pregnancy. Although not scientifically proven yet, I think there are observations that those who took a lot of herbs, the baby tends to have jaundice because their [babies'] livers just cannot cope with the toxics. So I don't take Chinese herbs.

By drawing on an interpretation of western scientific discourses, Chew Mee was able to 'legitimise' her rejection of various Chinese dietary rites commonly advocated for pregnant women.

In more ambivalent ways, Whee Hui (a manager in her late twenties) counters culturally imposed food prohibitions by exploiting other potentially contradictory strands in the same cultural discourse on pregnant bodies. Among the Chinese, pregnancy elevates a woman's status, as procreational ability is highly valued. A consequence of this is the view that certain normally 'irrational' demands such as food cravings should be not only tolerated but satisfied. While Whee Hui generally puts up with consuming cultural foods deemed to be *bu* (nourishing) and avoiding others believed to be undesirable, she asserts her privileged position to claim her right to a specific food craving:

> I like durians [a tropical fruit commonly deemed as 'very heaty' and hence taboo during pregnancy] a lot. Before getting pregnant ... I will eat durians, but not crave for them. But now [mid-way through her pregnancy], I *crave* for durians. I was so happy for the past few months because it's the durian season. Practically ... I tell you, every alternate day, my husband, my parents, my in-laws will have to go out and buy durians for me to satisfy my craving.

As Lim (1994: 12) puts it, '[food] cravings of pregnant women and people's acceptance of them may be a symbolic statement about the special dominant social position of pregnant women on account of their fulfilment of their procreational role.' By asserting her culturally sanctioned craving for a fruit that is usually culturally unacceptable for consumption by pregnant women, Whee Hui manipulates the cultural system to enlarge her own autonomy.

Other women resort to strategies of evasion, concealment and non-compliance to counter the culturally tinted inspecting gaze. A number of interviewees who do not live with their parents or parents-in-law admitted that they were far less vigilant about observing prohibition rites when they were out of range and beyond the supervisory eye of their mothers or mothers-in-law. In Wei Yuet's case, her mother-in-law, who lives in Malaysia, is a fervent believer in the benefits of Chinese herbs as *an tai yao* (i.e. medicine which will protect, comfort and strengthen the foetus within the pregnant woman) and unfailingly prepares this medicinal potion at every opportunity during Wei Yuet's fortnightly visits. Wei Yuet confessed that she had never once actually consumed the potion: each time, she would feign compliance, insist on bringing the medicine home with her to Singapore, and promptly pour it away on reaching home. In her words,

> I am very lucky that I am not staying with my mother-in-law ... because Chinese people believe that I have to drink [*an tai yao*] and when my mother-in-law gives it to me, I bring it home. But I won't drink it and she won't see it ... Then when I see her again, I just say that I did take those things.

Apart from negotiating culturally imposed dietary controls, pregnant women also have to contend with a number of other cultural rituals which have a disciplining effect on their socio-spatial movements. As with food taboos, while only 20.7 per cent of the women surveyed expressed outright disagreement with the traditional injunction that pregnant women must circumscribe their movements and be watchful of who and what they come into contact with in everyday life,[8] women also resort to a variegated range of strategies to inflect the cultural gaze.

Shu Min represents a minority of interviewees who claimed that they knew little about cultural taboos and could totally disregard them anyway:

> basically, I am a Christian, so I don't follow traditional customs very strictly … I don't know much, anyway. I only know that I am not supposed to knock things lah [sic], not supposed to paint, … my husband also. Both of us are not supposed to do a lot of taboo things … knocking, painting, drilling … But both of us don't bother. Like, when I was pregnant, we were renovating the house. They said cannot renovate house … cannot even go and see the renovation. But I still went!

Among other interviewees, some have been reminded of various spatial activities to be avoided during pregnancy but adopted strategies of concealment to cover their tracks. Felice was told

> not to watch movies that are too scary. Yeah … I know, even my mother tells me that although she is a Catholic as well. She just says that I should avoid such shows.

In response, Felice simply avoided the issue as far as possible:

> I love horror movies … I simply don't tell her [Felice's mother] to avoid any clashes.

Similarly, while Kristine had been forewarned about not moving the furniture around in her home while pregnant, she felt that practical considerations overrode these concerns and carried out her actions unobtrusively:

> We did quietly move the furniture. And it's all right. … because we need to decide where to put the cot. So we shifted things around to make room for the cot … but no one needs to know.

While there are instances of occasional non-compliance with culturally prescribed 'dos and don'ts', it would appear that, more so than state interventions or the medical gaze, cultural norms and forms as disciplinary practices exert a more consistently pervasive power over women's pregnant bodies in the sphere of everyday life. It is perhaps instructive to reflect on the intricate links between cultural discourses and ritual practice on the one hand, and strategies of control over the pregnant body on the other. The very moment a Chinese woman becomes pregnant, her socialisation of all kinds of traditional advice and cultural practices also begins, and persists at least until the end of pregnancy. Most of these cultural discourses and practices often hinge on the concept of 'good motherhood'. For a woman to be a 'good mother', she has a moral obligation to make sacrifices for the sake of her child(ren). Sacrifice does not take place only when the child is born, but begins at the very moment of conception. 'Good' motherhood means that a pregnant woman must eat or do whatever is culturally perceived as beneficial to the foetus, and must refrain from any action traditionally deemed to be harmful to the development and eventual birth of the baby she carries, regardless of personal preferences. Failure to subscribe to these social rules makes women 'bad' mothers, which is further vindicated if the baby is born with any defect or disease. Many pregnant women accept the notion of 'good motherhood', adhering to traditional advice on cultural practices in their everyday lives and undertaking surveillance of their own bodies for the sake of the baby, as well as others who have a vested interest in and attempt to speak on behalf of the unborn. Hence, for many women, conformity to the voice of tradition is often not a question of belief but an outcome of buying into cultural notions of what constitutes 'good' motherhood. As Choon Eng (an administrator in her late twenties) puts it,

> they are all old wives' tales, these dos and don'ts, but it's like, if they say don't do it and you don't and when the baby comes out with problems, there's nothing to say. But if you really did it … when it really happens … I mean sometimes it's just so coincidental, then you will feel guilty yourself. People have told you and you don't believe, and now, you see what happens? Just like my girl's birthmark. I did not paint when I was pregnant but she still has a birthmark. That is just natural, so nobody can say anything. But if someone had told me not to paint but I still purposely painted, and coincidentally she has a birthmark, I will feel guilty. [Furthermore] her birthmark is on her hand [and therefore outwardly visible] I will feel quite bad … Honestly speaking, we are more modern now, but it's still better for us to heed the advice given. Hear what they say and obey what they say. So when the baby is born safe, we will also be very happy … it's better to listen and follow what they say rather than regret later on.

## Conclusion

The exercise of power over pregnant bodies is instituted through an inspecting gaze which, while externally directed by the agents of state, medical and cultural authorities, gains effect when turned inward by the women themselves. As Foucault (1979: 155) writes:

there is no need for arms, physical violence, material constraints. Just a gaze. An inspecting gaze, a gaze which each individual under its weight will end by interiorising to the point that [s]he is [her] own overseer, each individual thus exercising this surveillance over, and against h[er]self.

In the same vein, Purushotam (1992: 328) argues that as power 'invests the body with its socially productive capacities ... These same capacities also, and at once, subject the body to power'. At the same time, it is clear that these dominant discourses and specific techniques of power are often punctuated and undermined because ideas of 'appropriate' management of pregnancy and behaviour of pregnant women are not always shared by all individuals. As Foucault (1980b: 162) puts it, 'there would always be ways of slipping through their net, or that resistances would have a role to play'. This chapter has considered the network of strategies that Singapore Chinese women take towards resisting and negotiating state, medical and cultural impositions as related to reproduction and pregnancy in general, and pregnant bodies in particular. We have done this by considering the body as 'a conduit which enables political agency to be thought of in terms of transgression and resistance' (Callard 1998: 387). While such agency, in our analysis, did not escalate to take on highly visible or collective forms, what is at work on the pregnant body is a 'microphysics of power' (Foucault 1979: 26) conceived of in terms of 'strategic positions in a constantly active, even battling network of relations' (Purushotam 1992: 329) mapped into the realm of the everyday.

The pregnant body, in so far as it is to be considered a political terrain where contestations of gender relations are played out, cannot be viewed solely in physiological terms. Instead, it must be interpreted in terms of prevailing gendered state, medical and cultural discourses and practices, as well as through the experiences of women in the spaces of everyday life. In these respects, the pregnant body holds particular salience. Feminists intent on challenging patriarchal notions of gender relations in society often have to contend with the dilemma of either claiming complete equality in a gender-neutral world, or demand recognition and support for *women's* roles and responsibilities. On the one hand, the former strategy, often formulated by resting on men's qualities and activities, risks incorporating women as 'lesser men' (Pateman 1992: 236). On the other hand, the latter strategy grounds feminism in sexual difference and risks gliding 'down the slope to separate spheres' (Philips 1992: 219), because the notion of women being different from men easily justifies the relegation of women to a different domain of social life and also to a different status from men. The pregnant body lies at the centre of this perplexing dilemma. If women claim to be the same as men, then this brings into question the whole range of differential social treatment and expectations of the pregnant mother – maternity leave, exemption from national military service, specialised medical care, and other tangible and non-tangible privileges in private and public spaces. If women claim that the pregnant body makes them fundamentally different from men, this may lead to the naturalisation and normalisation of different and unequal social positions for men and women. Emancipatory politics hence mean that there is no one single formula to contesting the multiple oppressions at work in women's lives; instead, as the women highlighted in this chapter have shown, power relations have to be engaged in myriad ways, at both the discursive and everyday, habitual levels. Clearly, then, the wide range of reflexive strategies women draw on testifies to the salience of viewing the pregnant body as a terrain of continuous power and resistance, and betrays the notion of the body as a passive, docile and intensely conservative subject.

## Notes

1   In a multi-ethnic society such as Singapore, there is a diversity of discourses and rituals surrounding the pregnant body among the different ethnic and cultural communities. In this chapter, we focus solely on Chinese women, not only because the Chinese constitute the majority ethnic group in Singapore but also, given our argument that the pregnant body is culturally inscribed, to be able to illustrate, with sufficient depth and integrity, the inner workings of one specific cultural discourse on the body.

2   The questionnaires were administered at a number of sites in Tampines New Town including a government polyclinic, private women's clinics, residences, shopping malls, transit points such as bus stops and train stations, at the foot of apartment blocks and other public spaces. A total of 403 questionnaires were returned, of which sixty-five were invalidated as a result of incompleteness or illegibility of handwriting.

3   The in-depth interviews were conducted either in English or Mandarin, depending on the interviewee's preference, and took place primarily in interviewees' homes. Each interview lasted between three-quarters of an hour to more than two hours. All interviews were taped and transcribed for analysis.

4   The state often exercises a form of power which Foucault (1982: 213–15) calls 'pastoral power' to emphasise the focus on reforming a people's health or habits.

5   This refers to the 'rational application of medical knowledge and practice to the production of healthy, reliable, effective and efficient bodies' (Turner 1992: 21).

6   Chinese cultural concepts relating to pregnancy basically revolve around the *yin–yang* spectrum which ranges from weakness to strength. Traditional Chinese understandings of health and illness rest upon the functions of the five *yin* 'organs' and six *yang* 'viscera' in human bodies. Blood and *ch'i*, the essences which are functionally linked via energy pathways, are seen as a paired *yin–yang* dualism. The health of individuals is a matter of balance between the various forces of *yin* and *yang*, qualified by unobstructed energy flow. Under this system, blood is believed to be the dominating aspect of the female constitution. In this system, pregnant bodies, which are subjected to 'more or less serious depletions of blood, ... make them chronically susceptible to the disorders accompanying such bodily loss' (Furth 1987: 13).

7   Prohibition rituals are of two types: avoiding foods considered too 'cooling' or too 'heaty', and avoiding foods based on the theory of signatures (for example, crabs, squid and other foods with too many limbs should be avoided as they will lead to hyperactive, fidgety babies).

8   Various everyday activities are deemed to harm the foetus or result in deformities in infants. They include sewing (especially when sitting on the bed), hammering nails, painting, moving pieces of furniture around, coming into physical or visual contact with 'ugly' or 'scary' animals (from apes in the zoo to Muppets on television!), attending 'red' (auspicious) events such as weddings as well as attending 'white' (inauspicious) events such as funeral wakes.

# References

Annandale, E. and Clark, J. (1996) 'What is gender? Feminist theory and the sociology of human reproduction', *Sociology of Health and Illness* 18, 1: 17–44.

Armstrong, D. (1995) 'The rise of surveillance medicine', *Sociology of Health and Illness* 17, 3: 393–404.

Bungar, J.B. (1991) 'Sexuality, fertility and the individual in Singapore society', unpublished MA thesis, Department of Sociology, National University of Singapore, Singapore.

Callard, F.J. (1998) 'The body in theory', *Environment and Planning D: Society and Space* 16: 387–400.

Crow, G. (1989) 'The use of the concept of "strategy" in recent sociological literature', *Sociology* 23, 1: 1–24.

Doran, C. (1996) 'Global integration and local identities: engendering the Singaporean Chinese', *Asia Pacific Viewpoint* 37, 2: 153–64.

Duncan, N. (ed.) (1996) *BodySpace: Destabilizing Geographies of Gender and Sexuality*, London and New York: Routledge.

Dyck, I. and Kearns, R. (1995) 'Transforming the relations of research: towards culturally safe geographies of health and healing', *Health and Place* 1, 3: 137–47.

Foucault, M. (1979) *Discipline and Punish*, New York: Vintage Books.

——(1980a) *The History of Sexuality Volume 1: An Introduction*, translated by R. Hurley, New York: Vintage Books.

——(1980b) *Power/Knowledge: Selected Interviews and Other Writings, 1972–77*, New York: Harvester Wheatsheaf.

——(1982) 'The subject and power: an afterword', in H. Dreyfus and P. Rainbow (eds) *Michel Foucault: Beyond Structuralism and Hermeneutics*, Brighton: Harvester.

——(1985) *The History of Sexuality Volume 2: The Use of Pleasure*, translated by R. Hurley, New York: Vintage Books.

Furth, C. (1987) 'Concepts of pregnancy, childbirth, and infancy in Ch'ing Dynasty China', *Journal of Asian Studies* 46, 1: 7–35.

Giddens, A. (1981) *A Contemporary Critique of Historical Materialism Volume 1: Power, Property and the State*, London and Basingstoke: Macmillan.

——(1986) *Central Problems in Social Theory: Action, Structure and Contradiction in Social Analysis*, Basingstoke: Macmillan.

——(1987) *A Contemporary Critique of Historical Materialism Volume 2: The Nation-State and Violence*, Basingstoke: Macmillan.

Graham, E. (1995) 'Singapore in the 1990s: can population policies reverse the demographic transition?', *Applied Geography* 15, 3: 219–32.

Gregson, N., Rose, G., Cream, J. and Laurie, N. (1997) 'Conclusions', in Women and Geography Study Group (ed.) *Feminist Geographies: Explorations in Diversity and Difference*, Harlow: Longman.

Hannah, M.G. (1997) 'Imperfect panopticism: envisioning the construction of normal lives', in G. Benko and U. Strohmayer (eds) *Space and Social Theory: Interpreting Modernity and Postmodernity*, Oxford: Blackwell.

Heng, G. and Devan, J. (1995) 'State fatherhood: the politics of nationalism, sexuality and race in Singapore', in A. Ong and M.G. Peletz (eds) *Bewitching Women, Pious Men: Gender and Body Politics in Southeast Asia*, Berkeley, CA: University of California Press.

Johnson, L. (1989) 'Embodying geography: some implications of considering the sexed body in space', *Proceedings of the 15th New Zealand Geographical Society Conference*, August, Dunedin.

——(1990) 'New courses for a gendered geography: teaching feminist geography at the University of Waikato', *Australian Geographical Studies* 28, 1: 16–27.

Kibria, N. (1990) 'Power, patriarchy and gender conflict in the Vietnamese immigrant community', *Gender and Society* 4, 1: 9–24.

Kuah, K.E. (1997) 'Inventing a moral crisis and the Singapore state', *Asian Journal of Woman's Studies* 3: 36–70.

Lefebrve, H. (1991) *The Production of Space*, Oxford: Blackwell.

Lim, Y.Y. (1994) 'Pregnancy and childbirth: interpreting the experience', unpublished academic exercise, Department of Sociology, National University of Singapore, Singapore.

Longhurst, R. (1997) 'Locating the biology of pregnant bodies', paper presented at the *Cultural Turns/Geographical Turns Conference*, 16–18 September, Oxford University, United Kingdom.

——(1998) '(Re)presenting shopping centres and bodies: questions of pregnancy', in R. Ainley (ed.) *New Frontiers of Space, Bodies and Gender*, London and New York: Routledge.

Martin, B. (1988) 'Feminism, criticism, and Foucault', in I. Diamond and L. Quinby (eds) *Feminism and Foucault: Reflections on Resistance*, Boston: Northeastern University Press.

Moss, P. and Dyck, I. (1996) 'Inquiry into environment and body: women, work, and chronic illness', *Environment and Planning D: Society and Space* 14, 6: 737–53.

Pateman, C. (1992) 'The patriarchal welfare state', in L. McDowell and R. Pringle (eds) *Defining Women: Social Institutions and Gender Divisions*, Oxford: Polity Press.

Philips, A. (1992) 'Feminism, equality and difference', in L. McDowell and R. Pringle (eds) *Defining Women: Social Institutions and Gender Divisions*, Oxford: Polity Press.

Phua, L. and Yeoh, B.S.A. (1998) 'Everyday negotiations: women's spaces and the public housing landscape in Singapore', *Australian Geographer* 29, 3: 309–26.

Purushotam, N. (1992) 'Women and knowledge/power: notes on the Singapore dilemma', in K.C. Ban, A. Pakir and C.K. Tong (eds) *Imagining Singapore*, Singapore: Times Academic Press.

Rose, G. (1993) 'Some notes towards thinking about the spaces of the future', in J. Bird, B. Curtis, T. Putnam, G. Robertson and L. Tickner (eds) *Mapping the Futures: Local Cultures, Global Change*, London and New York: Routledge.

Routledge, P. (1997) 'A spatiality of resistances: theory and practice in Nepal's revolution of 1990', in S. Pile and M. Keith (eds) *Geographies of Resistance*, London: Routledge.

Scott, J.C. (1985) *Weapons of the Weak: Everyday Forms of Peasant Resistance*, New Haven: Yale University Press.

——(1990) *Domination and the Arts of Resistance: Hidden Transcripts*, New Haven: Yale University Press.

*Straits Times* (2000a) 'Raft of ideas on how to trigger a baby boom', Singapore, 15 March.

——(2000b) 'Govt look to non-grads to boost births', Singapore, 16 March.

Teo, P. and Yeoh, B.S.A. (1999) 'Interweaving the public and the private: women's responses to population policy shifts in Singapore', *International Journal of Population Geography* 5: 79–96.

Turner, B.S. (1984) *The Body and Society: Explorations in Social Theory*, Oxford and New York: Blackwell.

——(1992) *Regulating Bodies: Essays in Medical Sociology*, London: Routledge.

——(1996) *The Body and Society: Explorations in Social Theory*, London: Sage.

Yeoh, B.S.E. (1995/6) 'The regulation of bodies: Singaporean men', academic exercise, Department of Sociology, National University of Singapore.

Yeoh, B.S.A. (1997) 'Sexually transmitted diseases in late nineteenth- and twentieth-century Singapore', in M. Lewis, S. Bamber and M. Waugh (eds) *Sex, Disease and Society: A Comparative History of Sexually Transmitted Diseases and HIV/AIDS in Asia and the Pacific*, Westport and London: Greenwood Press.

Yeoh, B.S.A. and Huang, S. (1995) 'Childcare in Singapore: negotiating choices and constraints in a multicultural society', *Women's Studies International Forum* 18, 4: 445–61.

# 3  Body politics in Bangladesh

*Tatjana Haque*

## Introduction

This chapter explores how low-income women in Bangladesh experience empowerment on a day-to-day basis. Emphasis is placed on women's agency in negotiating new identities, spaces and greater bodily self-possession. Before discussing the concepts of empowerment and agency, I will briefly sketch the general situation of women in Bangladesh and the role non-governmental organisations (NGOs) play in this context. The vocational training programme for women offered by the grassroots organisation Gonoshasthaya Kendra provides the background for my analysis of women's empowerment practices.

Although there is a plethora of case studies on empowerment projects in Bangladesh they tend to focus on the performance of the organisation rather than the subjects of the projects. This comes through in some of their titles: *In Quest of Empowerment: The Grameen Bank's Impact on Women's Power and Status*; *Breaking the Cycle of Poverty: The BRAC Strategy*; *Managing to Empower: The Grameen Bank's Experience of Poverty Alleviation*; *Transforming Women's Economies: BRAC*; *The Impact of Grameen Bank on the Situation of Poor Rural Women*.[1] Such a body of research provides an organisational outlook as well as a list of potential NGO strategies for empowerment, but fails to approach the issues involved in empowerment from the perspective of the women engaged in these organisations. Carr *et al.* (1996: 8) sums it up as 'put[ting] forward the voice of the people running the NGO, rather than the voice of the women who are the intended beneficiaries of their programmes'. Even Bhasin's (1985) and Batliwala's (1993) excellent work on women's empowerment derives from workshops with NGO leaders and not women participants. Along similar lines, Young (1993: 162–3) advises macro-level policy makers to 'consult' people about development goals and necessary resources, but refers them to NGOs in this context rather than as individual beneficiaries of NGO programmes.

My study argues for a reassessment of current notions of empowerment in such a way that women's own accounts of their lived and embodied experiences are given more emphasis. This approach to understanding empowerment acknowledges women's agency and ability to effect change in their own lives and in those of others.

## The situation of women in Bangladesh

In Bangladesh, as in several other South Asian countries, life expectancy for women is lower than for men. The sex ratio is ninety-four females for a hundred males (Atkins *et al.* 1997). A major reason for the masculine sex ratio in Bangladesh is the high rate of maternal mortality: 887 per 100,000 births (World Bank 1997). There is a gender discrimination in access to health care, education and training. Men and boys, when they fall ill, are more likely to receive medical attention than are women and girls (Kabeer 1989). Literacy rates for men and women in 1994 were 45 per cent and 24 per cent respectively (Bangladesh National Report to the Fourth World Conference on Women, Beijing 1995) and approximately 5 per cent of women were enrolled in technical and professional education (Mohila Parishad 1993). In addition, women's seclusion, sanctioned by traditional practices and beliefs, reduces mobility and limits women's employment opportunities (Kabeer 1989).

Since the mid-1970s, however, various social groups in Bangladesh such as the state, market institutions and NGOs have shown an interest in women due in part to international donor policies. For example, in 1976 the government declared the reservation of 10 per cent of government posts for women. In the same year, besides the Social Welfare Sector, a new sector was created for women, known as the Women's Affairs Division. While the Social Welfare Sector of the state continued to provide welfarist programmes for women, the Women's Affairs Division was tasked with the responsibility of promoting women's socio-economic status. In 1978, the Division was transformed into a fully-fledged Ministry, headed by the country's first woman minister. In Bangladesh, women have been/are heads of both the leading and the opposition parties. Even private market forces are said to have contributed to women's empowerment, particularly in the last decade. Ninety per cent of the 800,000 workers of the fairly new garment manufacturing industry in Bangladesh are women (United Nations Development Programme 1994). The majority of them are young and unmarried. While all these have contributed to changing women's status in Bangladesh, NGOs lead the way in promoting women's empowerment.

There has been a phenomenal rise in the number of NGOs since the inception of Bangladesh in 1971. In 1995, 16,000 NGOs were registered with the Directorate of Social Services, 1,600 with the Ministry of Women's Affairs and 850 with the Bureau of NGO Affairs (United Nations Development Programme 1994). Whereas in the early days NGOs were primarily involved in relief and rehabilitation work, they are now considered dominant actors in development intervention. Their activities include poverty alleviation, women's rights, education, health and family planning, and environment. They also play a major role in the current discourse on good governance and civil society. Influenced by recent gender and development debates and the fact that gender is in fashion among donors, many NGOs have identified women as their main target group. They reach approximately two out of the fifty-five million women in Bangladesh through projects such as credit and savings schemes, skills training, adult literacy, health education, family planning, legal awareness and consciousness-raising

(United Nations Development Programme 1994). While impressive, most of the NGOs with a gender focus have concentrated on generating income for women within sex-stereotyped activities such as livestock and poultry raising, kitchen gardening, food processing and handicrafts. Afsar (1990), for example, mentioned a survey undertaken by the Women's Affairs Department in 1986 in which 91 per cent of the more than 700 NGOs registered with the Women's Affairs Department were found to be involved in handicrafts programmes. There are, however, a few organisations which have applied innovative strategies of empowerment and which challenge the traditional gender division of labour. Their concern for women's position in society was not in response to recent donor influences, but existed already in their initial stages in the early 1970s. Gonoshasthaya Kendra (GK), which provides the organisational framework for my case study of women's empowerment, is one of them.

## Gonoshasthaya Kendra

In 1971, during the liberation war against Pakistan, a few young Bangladeshi doctors who were studying in the United Kingdom, among them Zafrullah Chowdhury, managed to mobilise money from Bangladeshi doctors all over the world to assist the freedom fighters (Ray 1986). These young men decided to return to Bangladesh to support the liberation war. They set up a 480-bed field hospital on the Indian border for the wounded (Chowdhury 1995). After the war, in 1972, the hospital transferred to Savar, 40 km north of the capital, Dhaka. In the early 1970s, Savar was a typical rural community without industry, a health complex or even an NGO. The post-liberation period was a euphoric time when '[y]oung leaders were thinking with vision and excitement about the possible future of their communities and became eagerly involved with the task of recon-struction after the devastation wrought by the war' (White 1991: 12).

GK was born out of this idealism. Gonoshasthaya Kendra – The People's Health Centre – is, as the name suggests, an organisation whose vision and prin-cipal objective is and always has been 'health for all'. Therefore all other projects are interwoven with this basic aim. GK approaches health from a holistic point of view and emphasises the complete well-being of a person and not just the absence of disease. It therefore addresses poverty issues such as malnutrition, illiteracy, lack of clean water and unsanitary living conditions. Concentrating on the poor, GK provided preventive and primary health care services for the surrounding villages where access to health services was almost non-existent.[2] Over the years, GK developed into a complex, integrated rural development project which included other sectors besides health, namely education, nutrition, agriculture, microbiology, vaccine research, herbal medicinal plant research, income generation and vocational training.

The vocational training centre for rural landless women was set up in 1973. The reasons for starting the vocational training centre were related to women's vulnerable position in society which was hindering their access to health services. The health needs of women and children were found to be inadequately met, because women depended on male family members to accompany them to the health centres and because they did not have their own money to pay for the services. In order to achieve a lasting impact through the health programme, GK had first to assure women's access to income and education.

Income generation through handicraft projects was believed to be most suitable for women in the post-liberation period. GK followed the trend and opened a jute handicrafts workshop in 1973. This was later recognised to be an unsatisfactory approach to women's employment because the internal market for jute handicrafts was saturated with products that appealed only to foreigners and westernised urbanites. Moreover, export was risky and unpredictable because of rapidly changing trends. In addition, the handicrafts approach did not offer women alternative roles. It simply reproduced gendered stereotyped images of women and homemaking. GK therefore expanded into new areas of work from 1976 onwards. Various gendered vocational skills were introduced in the following years: for example, metalwork (welding, lathe operation and sheet-bending) in 1976, carpentry and shoe-making in 1978, baking and catering in 1979, fibre-glass fabrication in 1982, blockprinting in 1984, professional driving in 1986, construction work in 1987, letter composition and printing in 1988, and irrigation pump operation, repair and maintenance in 1989.

GK places great emphasis on demonstrating that women can perform roles other than those ascribed to them by socio-cultural norms. It deliberately aims to challenge myths about what women can and cannot do. In addition to creating new employment opportunities, GK provides a space for women to establish collective social relationships. These offer women a source of social support which can complement or even replace family and kinship networks. To the outside observer who enters the GK premises for the first time, the organisation comes across as a self-sufficient independent commune. It offers hostels for the unmarried workers and residential quarters for the married couples, as well as a school and day-care centre for the workers' children. It also houses meeting and dining halls, a television room, a hospital offering health and family planning services, a tea stall, a bank and the various manual and technical workshops.

GK's approach towards women's empowerment is hence distinguishable from other NGOs in Bangladesh because of its pioneering role in training women in non-traditional manual and technical skills and in its physical arrangement, which is a commune-like and self-sufficient community. Both these characteristics make it an interesting vehicle for exploring women's empowerment practices. Having women perform in 'men's' jobs, wear 'men's' clothes and handle 'masculine' tools affects women's bodily comportment. They speak, move and express themselves through their bodies in new ways. The fact that non-family-related men and women work, sit, eat and live together in close proximity on a daily face-to-face basis is highly unusual. Eventually, though, women (and men) get used to this peculiar and unfamiliar situation and appreciate it. For many women the initially shocking environment at GK turns into a 'family environment', which allows women to lose their inhibitions towards men. However, before I enter a discussion on women's daily experiences of embodied empowerment, I will first introduce relevant theoretical conceptions of women's empowerment and agency.

## Moving towards an understanding of women's lived experiences of empowerment[3]

### Empowerment in the development discourse

Empowerment in the development context has its roots in Freire's conscientisation approach (1972). According to Freire's philosophy, all human beings, even if they are kept 'submerged' in the 'culture of silence', have the potential to reach a critical perception of reality and its contradictions. Through what he calls 'problem-posing education' and 'dialogical encounters with others', the 'oppressed' can become conscious of their own realities and react in a critical way. Feminists expanded this approach by incorporating a gender dimension that was lacking in Freire's analysis (Batliwala 1993). Empowerment became the key goal of feminist grassroots organisations. The empowerment approach emerged in the mid-1980s from a strong criticism of women's integration into mainstream 'Eurocentric' western-designed development and from disillusionment with the Women in Development (WID) approaches which had failed to acknowledge the possibility of women's 'double exploitation through class and gender relations' (Wood 1992: 16).[4]

Unlike conventional development projects, which tend to be bound within fixed time limits, feminists highlight the notion of process in empowerment:

> There is general unanimity that empowerment does not refer to an end-of-project product or state that can be attained within defined time-frames. Instead, empowerment is best understood as a dynamic and on-going process which can only be located on a continuum.[5]
>
> (Shetty 1991: 13)

Schuler (1986), for example, identifies various stages within the process of empowerment, moving from individual to collective consciousness-raising to mobilisation, in pursuit of political and legal change. In Sharma's (1991–2) model, the process of empowerment involves individual self-assertion, collective resistance and mobilisation that challenges power relations. Rowlands (1997) classifies the process of empowerment into three dimensions of personal empowerment, empowerment in close relationships and collective empowerment.

Although the feminist perception of empowerment appears to be that of a process rather than an end product, there is little in-depth analysis of how this process evolves. The literature is fairly abstract. It highlights various steps, stages or points within the process of empowerment (individual awareness, collective consciousness, collective action), but fails to capture the flow of this process. It does not elaborate adequately on how those stages are linked to each other. In other words, it lacks a deeper understanding of how women become empowered and what that means to them. How do women move from one position to the other within the empowerment process? What are they experiencing during the 'in-between' phases? Rowlands (1997) is one of the few who combines a thorough theoretical discussion of empowerment with its practical implications in her work on women's empowerment in Honduras. Most of the empowerment literature, however, describes the process of empowerment in theoretical and not in practical terms. Selected steps of empowerment are placed on an 'empowerment continuum' (Shetty 1991), where the final step is usually that of 'collective action'. 'Collective action' tends to be the ideal outcome of a feminist perception of empowerment and appears to have political connotations: '[E]mpowerment must entail as an ulti-mate goal the ability of the disempowered to act collectively in their own practical and strategic interests' (Kabeer 1994: 256).

What constitutes 'collective action'? The literature describes it through exam-ples, such as: struggling for fair wages; protesting against illegal divorce, domestic violence, polygamy and dowry practices; organising for the shut-down of local liquor shops; retaining control of forests; complaining about local government corruption; fielding candidates in local elections; and participating in national women's movements (Kramsjo and Wood 1992; Kabeer 1994; Westergaard 1994). 'Collective action' can supposedly be 'facilitated' through NGO interven-tion, usually through awareness-raising strategies, but how do women experience this 'facilitation'? How does the political awareness of an individual woman translate into 'collective action' of an entire group? 'Collective action' does not happen overnight. What do women experience that makes them decide to participate in collective political action?

In order to get a better grip on 'collective action', I believe one needs to step back and rewind the process of empowerment. One should spend some more time exploring the individual's daily lived and embodied experiences of empow-erment. Collective political action involves a certain degree of political awareness, but is nevertheless an 'action'. The notion of social change through 'collective action' requires people to 'act out' change and such actions require the physical use of the body.

## Bodies and empowerment

Recently the body has received increased attention in academic discourses[6] as well as in popular interest. Most of the literature, however, reflects on how people in the west perceive their bodies:

> In the affluent west, there is a tendency for the body to be seen as an entity which is in the process of becoming; a *project* which should be worked at and accomplished as part of an *individual's* self-identity.

> (Shilling 1993: 4–5)

The body is addressed from various angles. One current tendency in the west is to focus on reconstructing the body into a healthier, fitter and better-looking object. People are absorbed by 'competitive pressures of self-presentation' (Shilling 1993: 92) visible in the increased phenomena of anorexia and eating disorders, a recent obsession with fitness, slimness and plastic surgery. 'Medical', 'slender' and 'plastic' bodies are the outcomes of rigid health and beauty regimes (Bordo 1993). New means of telecommunications create 'technobodies' or 'virtual bodies' – bodies that are now able to 'act at a distance' (Pile and Thrift 1995: 7; Callard 1998). Feminist discourses centre around 'sexed bodies' in their deconstruction of the 'masculinist claims to knowledge' which allow white men – but not others – to transcend their embodiment by perceiving their bodies as the containers of consciousness (Longhurst 1995: 98). Recent feminist theories therefore challenge these 'disembodied' concepts of reason, arguing that the mind is not separable from the body: '[o]ur dominant conceptions and ideals of reason have been connected to bodies ... [and] have been expressions of bodily concerns or needs and reflections of embodied ways of being' (Alcoff 1996: 17).

Western philosophical and political binaries such as mind/body, sex/gender, male/female are being contested (Grosz 1994; Longhurst 1995). One speaks of 'third bodies' and 'cyborg bodies', where the body takes on the characteristics of a machine (Haraway 1991). Queer theories expand on 'hybrid bodies' (Bell *et al.* 1994; Callard 1998). 'Raced bodies' are the subject of analysis in post-colonial theories concerned with the relation between colonial and colonised bodies (Spivak 1985).

My approach towards the body, on the other hand, is to reach a better understanding of women's embodied experiences of empowerment in the context of Bangladesh. In Bangladesh, however, the body is an area of research to which neither empowerment nor development debates have given any particular significance. In mainstream development literature, women's bodies were either reduced to human resources for productive purposes or targeted as means to implement population control. In her bibliography on research on women in Bangladesh, Islam (1994) identified the following gaps in the literature: feminism and the women's movement; feminist methodology; women in patriarchy; structural adjustment and women; and women and social change. One finds no reference, however, to how the process of change affects women's bodies, how women's bodily landscapes become transformed and how women practise empowerment through their bodies. Even in the wider regional context of South Asia, feminist scholarship has failed to incorporate an understanding of women's bodies that goes beyond issues of production, reproduction and sexual oppression (e.g. the maternal body, the body as bearer of women's double burden, the victimised body). Recent work on women's empowerment has begun to challenge the notion of women as victims. However, when speaking of empowerment and collective action, feminists – as discussed above – seem to focus on the mind and consciousness, ignoring the physical aspects of women's empowerment process: '[T]he process of empowerment begins in the mind, from woman's consciousness ... from believing in her innate right to dignity and justice, and realising that it is she, along with her sisters, who must assert that right' (Batliwala 1993: 9).

This approach envisions a certain outcome – i.e. women fighting collectively for their rights – but fails to explain how change manifests itself in women's bodies. When women experience empowerment they do this through their bodies in both mental and physical terms. Empowerment, however, is a process, not an end in itself. To be able to think of possible outcomes of empowerment, one needs to first ask the women themselves how they experience processes of change in and through their bodies. For a better understanding of women's lived and embodied experiences of empowerment, I draw on a specific strand of literature: that concerned with psychoanalytical elements of social behaviour. I find Giddens' (1984) theory of structuration and Goffman's (1959, 1963) work on 'interaction practices' and people's behaviour in public particularly useful.

### Agency, spatiality and the body

In his theory of structuration, Giddens (1984) perceives people as active agents. He refers to them as 'knowledgeable agents', capable of reflexively monitoring their actions. They know about the 'conditions and consequences of what they do in their day-to-day lives' (Giddens 1984: 281). Giddens, however, does not separate agency from structure. In his model, structures (sets of rules and resources) are perceived as both constraining and enabling: 'Each of the various forms of constraint are thus also, in varying ways, forms of enablement. They serve to open up certain possibilities of action at the same time as they restrict or deny others' (Giddens 1984: 173–4).

Human agency does not take place in a void – in a vacuum. In Giddens' model it is 'contextually situated' within temporal and spatial frameworks. According to Giddens, individuals are positioned in daily 'time–space paths' as well as within their 'life paths'. These positions can be within various zones ranging from home, work, neighbourhood, city and nation to worldwide systems (Giddens 1984: 85). Giddens refers to these places as 'locales', which he defines as 'settings of time–space paths through which individuals move' (Giddens 1984: 367).

Agents, however, do not act in these different time–space paths just as individuals *per se*. They also act towards others and are acted upon by others. People meet in various 'locales'. Each space will have its own influence on the type of interaction that can take place between people. It will also have an impact on people's bodily conduct and bodily space. These in themselves are gendered in terms of different masculine and feminine ways of using the body and experiencing space (Young 1989).[7]

Erving Goffman's (1959, 1963) work on interaction, institutions and the 'positioning of the body in social encounters' plays a significant part in Giddens' structuration theory. Goffman's starting point for analysis is not the individual agent, but different forms of interaction between people. Interaction, in terms of face-to-face interaction, is defined by Goffman as 'the reciprocal influence of individuals upon one another's actions when in one another's immediate physical presence' (1959: 26). He distinguishes between different types of interactional occasions. 'Focused gatherings', for example, require the focused attention of the participants (for example, people engaged in conversation, dancing couples, pairs of co-operating workers), whereas 'unfocused gatherings' can include, for example, pedestrians meeting in the street (Kendon 1988). However, according to Goffman, even in 'unfocused' situations like street encounters, certain 'traffic rules' of non-verbal face-to-face interaction take place (people glancing at each other, acknowledging each other's physical presence while passing each other):

> Goffman's most telling contributions to understanding the sustaining and reproduction of encounters are to do with the relation between the reflexive control of the body – that is to say, the reflexive self-monitoring of gesture, bodily movement and posture – and the mutual co-ordination of interaction.
> (Giddens 1984: 78)

This reflexive monitoring of action demands the 'exhibiting of presence' (Goffman 1963) – in other words, awareness of how one dresses, moves and appears in public. People manage their bodies according to the spaces in which they move. Goffman draws a link between embodied action and people's perception of their social surroundings. His approach refers to people's 'understanding of the (social) situational accommodation which any average stroll entails' (Crossley 1995: 136).

> [Goffman] is not only concerned with *what is done to the body* in the context of the social world, the ways in which *it is acted upon* and *represented*, he is concerned with *what the body does* in the social world, *how it works to construct and reproduce that world*, how *it acts*. The body acts and is acted upon for Goffman. It sees and is seen, speaks and is spoken to and about.
> (Crossley 1995: 147–8, my italics)

## Embodied agents of change

In this section I describe how I have translated the theoretical insights of Giddens' and Goffman's work into my practical research agenda. I do this by investigating how the experience of leaving their homes, becoming involved in an organisation like GK and interacting with others transforms women into embodied social agents of change. Applying Giddens' theory to women's empowerment practices I would argue that empowerment is a process influencing and reshaping women's lives, and organisations like GK are examples of 'locales' where empowerment practices are likely to take place. The 'locale' GK can be an enabling structure in terms of providing women with income, skills, literacy, information and space to interact with others. However, it can also be constraining through its different rules and regulations. In fact, questions are being raised about the extent to which NGOs in Bangladesh are patronising and interfering in people's private lives (Devine 1996). This, of course, is tied into the way wider social structures function in Bangladesh. Village life, for example, can be very restrictive for women: limiting women's mobility; controlling access to education; and expecting particular behavioural codes from women. At the same time these structures can also be enabling in terms of offering a familiar surrounding, kinship networks and protection.

While accepting Giddens' general framework about the relationship between structure and agency, in this essay I am particularly interested in the process of agency. In concrete terms, my focus lies on what women actually do within given structures, what they make out of an organisation like GK, how they use the enabling resources for themselves, how they act within given constraining structures, and how they experience change. There is little in the development literature on Bangladesh which looks at how women experience processes of change. This constitutes a research paradox which at least one commentator has already noted:

> While not wishing to deny the systematic disadvantages which face women, especially rural and poor, this paradox arises from the way the initial questions are asked about women. Much commentary proceeds from the question 'How are women constrained?' instead of asking 'What do women do?' The first treats women, *a priori*, as passive (intransitive); the second as active (transitive).
>
> (Wood 1992: 13)

During my fieldwork period in 1994–5, I lived as a participant observer in the organisation's hostel together with the women workers for approximately seven months. I saw them daily, ate with them in the same canteen and participated in various meetings, cultural events and demonstrations organised by GK. Applying a multi-method approach, I undertook a questionnaire survey with approximately 80 per cent of the women engaged at the vocational training centre (127 women) to gain a socio-economic profile. My core method was ethnographic semi-structured in-depth interviews with thirty women, some living and working in the organisation, others commuting from their villages. Furthermore, I conducted contextual interviews with family members, non-working village women, garment workers and their employers, GK staff and directors, staff from other NGOs, government officials, academics and activists from the women's movement.

After repeated listening to what the women told me about their daily experiences of change and my own reflections on their accounts, I became more and more drawn to an aspect of empowerment which the literature does not address: 'embodied empowerment'. Whereas the literature tends to focus on economic empowerment in terms of an increased economic well-being via income generation or political empowerment via consciousness-raising, I wish to highlight the embodied dimension of empowerment in this chapter. Elsewhere I have argued for a reassessment of current notions of empowerment by paying more attention to embodied aspects of empowerment regarding women's changed body language (for example, voice, dress, movement, posture) and behaviour towards others (Haque 1997, 1999a, 1999b).[8] Here I am drawing on a selection of examples that will illustrate women's new appearance and self-presentation and how this affects women's agency and interaction with others. Through the example of selected case studies I will explore how women 'act and are acted upon', 'see and are seen', 'speak and are spoken to' in their new bodies.

### Individual action

Majilla[9] came to GK in 1992 and became involved in the carpentry training section. At the time of our interviews in 1994, she was 25 years old and lived with her husband and two children in a village nearby. They had left their original home village in 1990 and moved closer to the capital to secure the family's well-being. Her husband now works in a brick factory. Both children go to GK school. Majilla perceived her former life in one of the more remote villages in Bangladesh as limiting and decided to leave as she could not see any future for herself and her family. She had heard about GK's special training programme for women and came with an interest in acquiring new skills. Her process of change, however, had already begun before she joined GK. She had been to a night-school where she learnt how to read and write. She also learnt from her observations of village life where some of the women had been oppressed by their husbands. She is continuing her learning process at GK:

> [We learn] about many things at GK, about the body and our health ... to drink water from the tubewell ... They tell us about certain medicines and advise us about child care.

> *Interviewer:* Do they talk about the situation of women in Bangladesh?
> *Respondent:* Yes. There is a lot of talk, for example, about husbands treating their wives badly or abandoning their wives ...
> *Interviewer:* Do the women talk about all these problems?
> *Respondent:* Yes, they talk about everything. You know I feel really good about that.
> *Interviewer:* Do they [GK staff] give any suggestions?
> *Respondent:* Yes, they give advice ... They organise special meetings for women once or twice a month.

She added:

> In my village many women got beaten up by their husbands. Now when I go to the village I console them and explain things to them. They say to me, 'Sister, you have done a very good thing by leaving this village.' So I have already learnt some things before [coming to GK], but from here [GK] I've learned a lot more.

Mimosa (aged 25) has also experienced oppression – in her case her own mother's:

Whatever my father dictated my mother had to do. He beat her often … but now the atmosphere has changed. Girls of today can work. They can say, 'We will work to support ourselves and won't tolerate being beaten … Why should we tolerate so much scolding and beating?' But for them [mothers] it's not like that. They manage all household chores and still tolerate all the beatings … My father punished my mother, scolded her and swore in our presence … he beat my mother so much [that's when I thought] I'd rather go out and work, I won't be beaten like my mother.

Mimosa came to work in order to avoid a similar situation to her mother's. She started her training at GK in 1987 and became involved in various sections such as welding, construction work and now jute plastic production. She lives with her son at the home of her parents in a nearby village. She was married at the age of fifteen, but her husband abandoned her and took a second wife. Mimosa learnt how to read and write at GK. Besides having more control over her life since joining GK, she also feels less of a burden to her family. On the contrary, she now contributes significantly to the family income:

Before, my father said, 'She is living on us, we have to provide food and clothes … to marry her off again costs a lot of money.' … Now he realises that I earn for myself and he will not have to take any burden.

The above quotes illustrate how GK operates as a springboard for women, a place where women come, obviously with their own baggage of knowledge and experience, but where their thinking and transformation process is stimulated. Of utmost importance to the women is also the fact that GK provides them with a space to meet and interact with other women. I have been told repeatedly how talking with others, exchanging ideas, learning from each other, consoling and giving each other advice have become a significant part of their lives. They specifically highlighted how in their villages they did not have the same kind of friendships and relationships of trust, the same opportunities to be comforted when necessary. Given the cultural constraints on women's mobility in the context of Bangladesh, women do not have the same access to community-based networks as men (Kabeer 1994).

When I asked Majilla where the women at GK would go to for consolation she told me of her colleague, who came to her for advice and comfort:

Ramiza works with me. She has some education, whereas her husband has graduated. Still he sits idle at home. He does not work. She said to me, 'What shall I do? He does not work at all.' I said to her, 'He should go to work. What's the point of sitting idle after having taken so much education?' Yesterday her husband went with my husband to take the job of a digger.

*Interviewer:* So after you talked to him he started to work?
*Respondent:* Yes …

*Interviewer:* So he listened to you? He understood what you were talking
about?
*Respondent:* Certainly he understood.
*Interviewer:* Does this happen often, that women here help each other when
they have problems with their husbands?
*Respondent:* Yes, and I want everybody to live happily and help each other.

As can be seen from the above quote, Majilla engaged in an action that was
not related to her own or her family's well-being, but to a friend's, her work
colleague. She was confronting her colleague's husband and thereby challenged
existing power relations. Not only was this man a stranger to her, but he also
represented a different class, that of an educated person. She was not intimidated by that fact and even managed to have an influence on him.

Women like Majilla do not only see themselves as being different from before.
They also differentiate themselves from other women:

GK makes women strong ... women from my village would not be able to start
to work at a garment factory [because] they are scared of unknown people.

She also reported that:

my mother can't talk to people. She is also scared to get in a car. She doesn't
know what it is like to do a job like mine ... When she walks in the street she
covers her face with a veil. She doesn't look at people's faces properly.

Working at GK requires women to move around and 'expose' their bodies in
front of non-familial male members. Women have to interact with these men on
a daily face-to-face basis. The work situations at GK are examples of what
Goffman (1959, 1963) describes as 'focused gatherings' in which the women have
to apply 'interaction practices' with others. After having gone through a period
of adaptation, they begin to feel comfortable with those with whom they work
on a daily basis. This again is reflected in their bodily behaviour and movements.
The confident appearance they have gained radiates through their bodies in the
way they look at others, the way they hold their bodies, the way they move, the
way they talk to others, the way they themselves feel about and relate to their
own bodies.

This also makes them feel more at ease than other women when handling
encounters in 'unfocused gatherings': for example, on the streets on their way to
work. They no longer feel ashamed of moving in public spaces. They walk with
an air of confidence and talk back to any stranger if the situation requires it.
Majilla felt comfortable in arguing with her colleague's husband because she did
not feel intimidated by him:

I used to be scared before, but now I don't fear anybody any more – I really
don't.

Mimosa's experience was similar to Majilla's:

> Previously I never talked to other people, only my parents and grandparents. … I felt scared to talk to others … but now I move about. I speak about so many kinds of things with so many different people. I have changed in many ways [self-satisfied laughter] … Although I haven't had much education, I do know how to handle people now, I know how to talk to them. But before, if I spoke up they'd say, 'She doesn't know how to talk.'

Some women at GK carry this newly won confidence and speaking power to their villages. They have learnt to make themselves visible and heard. They have adopted a new and stronger position *vis-à-vis* the people they once feared and have succeeded in establishing a respectable identity within their own communities. They are challenging hierarchical structures through their behaviour and actions: they talk back and resist conventional norms. Through their behaviour most of them have managed to have an influence on others. They have, for example, become role models for other women in their villages. They give advice, sort out problems or find work for them. Mimosa has had an influence on her sisters:

> Now they also want to learn some vocation like me and follow my example … I went to many places to find jobs for my sisters.

Her community also seeks her out for her knowledge in family planning and health matters. Like many others she has become associated with GK – the health organisation – and therefore labelled as 'the medical woman'. Women like Mimosa hence serve as a link between the organisation and the communities, offering the latter a service:

> If someone has an ulcer or abscess I tell them to wash it with warm water, grind neem leaves and raw tangerine and apply it on to the affected area. Then the mats, bed sheets, quilts and own clothes must always be clean. I tell them these things … They know that I work in GK and take health classes. People come over to take suggestions from me.

However, agency and embodied aspects of empowerment are not only limited to individual women's relations to themselves and their interaction with others. They can also encompass, for example, the physical impact an entire group of women has on society in women's effort to claim public spaces.

### Collective action

NGOs like GK organise women to participate in national women's marches. The women join in because they are expected to do so. However, this does not mean that they are unaware of the effects this will have on the public. For many women, such occasions are often the first time they participate in political campaigns. They are conscious of their strength in numbers. They know that each of their individual bodies becomes part of a bigger body and that this will have an effect on the physical landscape of public spaces such as cities. Most women at GK also genuinely enjoy the marches. The feeling of 'togetherness' they have established at GK makes them feel comfortable when appearing in public as a collective. The friendships and relationships of trust women have developed at GK serve as a profound basis for creating notions of solidarity, not only among themselves but also with women beyond the immediate GK boundaries:

> It's really a great thing [the women's march]. We all do something together … we enjoy ourselves … we protest against men … I remember a girl from my village who got married but whose parents could not pay the dowry. A few days later the girl got killed. Nowadays less such incidents happen. Things are improving.
>
> (Majilla)

Various meanings are read into the marches. First of all, they stand for a collective demonstration of equality between men and women. In the case of the GK women, equality is understood in terms of their experience of being able to work just as well as men. Bringing the women's issue out into public through a very big and visible collective of women seems significant to the GK women: 'Everybody can see us on the road.' The marches are also interpreted as a protest against male dominance and violence against women. Women go there to claim their equal rights and their 'freedom'. The women's movement is described by the majority of GK women as beneficial for themselves and for future generations:

> Men shouldn't beat or abandon women. If we go on these marches and it will not help my generation, at least it will help my daughter's. That's why we need it.
>
> (Hanufa, aged 20, married, working in GK as a carpenter since 1985; her husband is a farmer, her daughter is at GK's child care centre)

As another woman put it:

> Women cannot move freely, they don't get their proper rights. I go to fight for these rights … I go every year. I never miss it. I like going.

> *Interviewer:* Is there a need for this movement?
> *Respondent:* Of course there is, because without such an initiative things won't change. We are doing it for a better future.
>
> (Zohra, aged 23, married, working in GK's printing press since 1987; her husband is enrolled in engineering studies. She was pregnant at the time of the interview.)

Participating in women's marches can therefore be a new way of experiencing political agency, something which was not possible for most women before. Such occasions have the potential of transforming women's individual agencies and commitments into collective action.

## Conclusion

This chapter sets out to challenge the representation of women in Bangladesh as a victimised, powerless and invisible group. It does this by exploring women's lived and embodied experiences of empowerment. My study highlights women's personal transformation to greater autonomy through their engagement with GK, both at work and in their social lives. Women are adopting new lifestyles, reinventing their relationships with others and reacting to their surroundings in a critical way. They have gained a broader understanding of the world that transgresses the confines of home and village boundaries. They see and present themselves in new ways leading to a redefinition of their appearance in private and public spaces. Women have acquired identities other than those of being a mother, wife or daughter. They have taken on additional roles, such as workers, members of an organisation, part of a collective, local healers and experts in health, and, last but not least, participants of a wider community, i.e. that of the national women's movement. Women's agencies are thus effecting change in their own and other people's lives.

However, the transition from these individualised transformations to collective mobilisation is still in progress. In GK, women have most certainly developed a sense of belonging, a sense of solidarity, a collective consciousness and a collective identity which all carry a strong potential for collective action. Collective action, as envisioned in the feminist discourse, where women collectively fight against discrimination and injustice and actively participate in decision-making is, however, not yet evident. The GK women are not taking on positions of political decision-making, they are not forming their own local organisations and are not organising themselves as a group to change local power relations. The type of collective action I have observed is that of participation in the national women's marches. Such action, however, is not taken from the women's own initiative as a collective, but is organised by GK. Does this imply that collective empowerment is being bestowed upon the women, or are external change agents like GK merely providing women with enabling tools for their process of empowerment? Does collective empowerment not have to emerge from the women themselves – in other words, come from within?

I think one needs to situate the notion of empowerment within a historical and geographical context. In Bangladesh, where cultural norms of female seclusion have kept most women from the public arena in the past, women's rallies – even if organised by NGOs – are definitely radical gestures of change. The collective bodily empowerment of women through street marches has the potential of transforming gendered landscapes of Bangladesh.

What then are the implications of GK's empowerment programme for patriarchal Bangladesh? GK perceives its vocational training programme for women as an experiment and a demonstration case to be copied by other NGOs or governmental projects on a wider national scale. The organisation's scope for reaching beyond its own boundaries is therefore clearly limited. Part of the problem, however, is also GK's lack of emphasis on creating local women's groups. GK's strategy involves drawing women from various villages to the NGO, rather than organising women's groups in selected villages. The result is an absence of collective mobilisation at the local level. The individual women's influences on their communities and their personal strategic actions and choices are on a small-scale and rather subtle level, but would have a strong potential for changing local power relations if they were multiplied through group formation. Women's bodily practices of empowerment are of particular significance in this context, as not enough attention has been paid to this area in the gender and development debate. Through their bodily expressions of empowerment, GK women are consciously and overtly resisting cultural norms of appropriate behaviour. I therefore find it timely and crucial to include an analysis of the body into the gender and development discourse. The implication of this must be to achieve a deeper understanding of how development intervention, directed towards women's empowerment, affects women's bodies, and to use this as a basis for strategic gender-sensitive development policy.

## Notes

1   Among others see, for example, Rahman (1986), Lovell (1992), Mizan (1994), Holcombe (1995) and Selim (1996). For a detailed critique of mainstream research on women in Bangladesh, see, in particular White (1992: 16), who argues that 'virtually every text on women in Bangladesh has been funded by foreign aid' aimed at the production of quantifiable evaluation results.
2   Hours and Selim (1989) quote the figure of one doctor to 30,000 people.
3   For a detailed theoretical discussion on empowerment, see Kabeer (1994), Rowlands (1997) and Haque (1999a, 1999b).
4   For a more detailed discussion on the WID approaches (welfare, equity, anti-poverty and efficiency), see Goetz (1991) and Moser (1991).
5   I find the term 'continuum' slightly ambiguous, however, as it seems to imply that change moves in a particular direction. I understand empowerment instead as a dynamic, non-linear process.
6   See, for example, among many others, Foucault (1977), Butler (1990), Fuss (1990), Featherstone *et al.* (1991), Bordo (1993), Shilling (1993) and Grosz (1994).
7   Young (1989) argues, for example, that women use only part of the body when engaged in an action, whereas men have the tendency to stretch their bodies and invade surrounding space. She also mentions different experiences of 'enclosed' and 'open' spaces and different ways of being positioned in space.
8   Among the many expressions of empowerment discussed in my doctoral dissertation the notion of 'embodied empowerment' plays a central role.
9   This – as well as the names of the other women quoted – is a pseudonym.

# References

Afsar, R. (1990) *Employment and Occupational Diversification for Women in Bangladesh*, International Labour Organisation (ILO) Asian Regional Programme for Employment Promotion (ARTEP) Resource Paper II, New Delhi: ARTEP.

Alcoff, L.M. (1996) 'Feminist theory and social science: new knowledges, new epistemologies', in N. Duncan (ed.) *Body Space*, London: Routledge.

Atkins, P.J., Kumar, N., Raju, S. and Townsend, J.G. (1997) 'Where angels fear to tread? Mapping women and men in India', *Environment and Planning A* 27, 12: 2207–15.

Bangladesh National Report to the Fourth World Conference on Women, Beijing (1995) *Women in Bangladesh: Equality, Development and Peace*, Dhaka, Bangladesh: Ministry of Women and Children Affairs.

Batliwala, S. (1993) *Empowerment of Women in South Asia: Concepts and Practices*, New Delhi: Food and Agriculture Organisation–Freedom for Hunger Campaign/Action for Development (FAO–FFHC/AD).

Bell, B., Binnie, J., Cream, J. and Valentine, G. (1994) 'All hyped up and no place to go', *Gender, Place and Culture* 1: 31–47.

Bhasin, K. (1985) *Towards Empowerment: Report of an FAO–FFHC/AD South Asia Training for Women Development Workers*, New Delhi and Rome: FAO–FFHC/AD.

Bordo, S. (1993) *Unbearable Weight: Feminism, Western Culture, and the Body*, Berkeley: University of California.

Butler, J. (1990) *Gender Trouble*, London: Routledge.

Callard, F.J. (1998) 'The body in theory', *Environment and Planning D* 16, 4: 387–400.

Carr, M., Chen, M. and Jhabvala, R. (1996) (eds) *Speaking Out: Women's Economic Empowerment in South Asia*, London: Intermediate Technology.

Chowdhury, Z. (1995) *The Politics of Essential Drugs: The Makings of a Successful Health Strategy: Lessons from Bangladesh*, London: Zed Books.

Crossley, N. (1995) 'Body techniques, agency and intercorporeality: on Goffman's relations in public', *Sociology* 29, 1: 133–49.

Devine, J. (1996) *Non-Governmental Organisations (NGOs): Changing Fashion or Fashioning Change?* Centre for Development Studies Occasional Paper 2, Bath: University of Bath.

Featherstone, M., Hepworth, M. and Turner, B. (1991) *The Body: Social Process and Cultural Theory*, London: Sage.

Foucault, M. (1977) *Discipline and Punish: The Birth of the Prison*, London: Allen Lane.

Freire, P. (1972) *Pedagogy of the Oppressed*, Harmondsworth: Penguin Books.

Fuss, D. (1990) *Essentially Speaking: Feminism, Nature and Difference*, London: Routledge.

Giddens, A. (1984) *The Constitution of Society*, Cambridge: Polity Press.

Goetz, A.M. (1991) 'Feminism and the claim to know: contradictions in feminist approaches to Women in Development', in R. Grant and K. Newland (eds) *Gender and International Relations*, Buckingham: Open University.

Goffman, E. (1959) *The Presentation of Self in Everyday Life*, New York: Doubleday.

——(1963) *Behaviour in Public Places: Notes on the Social Organisation of Gatherings*, New York: The Free Press.

Grosz, E. (1994) *Volatile Bodies: Toward a Corporeal Feminism*, Bloomington: Indiana University Press.

Haque, T. (1997) 'New identities for women in Bangladesh', in W. van Schendel and K. Westergaard (eds) *Bangladesh in the 1990s: Selected Studies*, Dhaka, Bangladesh: University Press.

——(1999a) 'Women, NGOs and "embodied empowerment"', paper presented at the *NGOs in a Global Future Conference*, 10–13 January, Birmingham, University of Birmingham.

——(1999b) 'Lived experiences of empowerment: a case study of a vocational training programme for women in Bangladesh', unpublished PhD thesis, University College London.

Haraway, D. (1991) *Simians, Cyborgs, and Women: The Reinvention of Nature*, New York: Routledge.

Holcombe, S. (1995) *Managing to Empower: The Grameen Bank's Experience of Poverty Alleviation*, London: Zed Books.

Hours, B. and Selim, M. (1989) *Une Entreprise de Developpement au Bangladesh: Le Centre de Savar*, Paris: L'Harmattan.

Islam, M. (1994) *Whither Women's Studies in Bangladesh?*, Dhaka, Bangladesh: Women for Women.

Kabeer, N. (1989) *Monitoring Poverty as if Gender Mattered: A Methodology for Rural Bangladesh*, IDS Discussion Paper 255, Sussex: IDS.

——(1994) *Reversed Realities, Gender Hierarchies in Development Thought*, London: Verso.

Kendon, A. (1988) 'Goffman's approach to face-to-face interaction', in P. Drew and A. Wootton (eds) *Erving Goffman: Exploring the Interaction Order*, Oxford: Blackwell.

Kramsjo, B. and Wood, G.D. (1992) *Breaking the Chains: Collective Action for Social Justice among the Rural Poor in Bangladesh*, London: Intermediate Technology.

Longhurst, R. (1995) 'The body and geography', *Gender, Place and Culture* 2, 1: 97–104.

Lovell, C.H. (1992) *Breaking the Cycle of Poverty: The BRAC Strategy*, Dhaka, Bangladesh: University Press.

Mizan, A.N. (1994) *In Quest of Empowerment: The Grameen Bank Impact on Women's Power and Status*, Dhaka, Bangladesh: University Press.

Mohila Parishad (1993) *Commission Report*, Dhaka, Bangladesh: Mohila Parishad.

Moser, C.O.N. (1991) 'Gender planning in the Third World: meeting practical and strategic needs', in R. Grant and K. Newland (eds) *Gender and International Relations*, Buckingham: Open University.

Pile, S. and Thrift, N. (1995) 'Introduction', in S. Pile and N. Thrift (eds) *Mapping the Subject: Geographies of Cultural Transformation*, London: Routledge.

Rahman, R.I. (1986) *Impact of the Grameen Bank on the Situation of Poor Rural Women*, Dhaka, Bangladesh: Bangladesh Institute of Development Studies.

Ray, J.K. (1986) *Organising Villagers for Self-Reliance: A Study of Gonoshasthaya Kendra in Bangladesh*, Calcutta: Orient Longman.

Rowlands, J.M. (1997) *Questioning Empowerment: Working with Women in Honduras*, Oxford: Oxfam Publications.

Schuler, M. (ed.) (1986) *Empowerment and the Law: Strategies of Third World Women*, Washington DC: OEF International.

Selim, G.R. (1996) 'Transforming women's economies: BRAC', in M. Carr, M. Chen and R. Jhabvala (eds) *Speaking Out: Women's Economic Empowerment in South Asia*, London: Intermediate Technology.

Sharma, K. (1991–2) 'Grassroots organisations and women's empowerment: some issues in the contemporary debate', *Samya Shakti* 6: 28–44.

Shetty, S. (1991) *Development Project in Assessing Empowerment*, Society for Participatory Research in Asia Occasional Paper Series 3, New Delhi: Society for Participatory Research in Asia.

Shilling, C. (1993) *The Body and Social Theory*, London: Sage.

Spivak, G.C. (1985) 'Three women's texts and a critique of imperialism', *Critical Inquiry* 12, 1: 243–61.

United Nations Development Programme (1994) *Report on Human Development in Bangladesh: Empowerment of Women*, Dhaka, Bangladesh: United Nations Development Programme.

Westergaard, K. (1994) *People's Empowerment in Bangladesh – NGO Strategies*, Centre for Development Research (CDR) Working Paper 94.10, Copenhagen: CDR.

White, S.C. (1991) *Evaluating the Impact of NGOs in Rural Poverty Alleviation. Bangladesh Country Study*, Overseas Development Institute (ODI) Working Paper 50, London: ODI.

——(1992) *Arguing with the Crocodile. Gender and Class in Bangladesh*, London: Zed Books.

Wood, G.D. (1992) 'Part One: introduction', in B. Kramsjo and G.D. Wood (eds) *Breaking the Chains: Collective Action for Social Justice among the Rural Poor in Bangladesh*, London: Intermediate Technology.

World Bank (1997) *World Development Report 1997*, Washington DC: World Bank.

Young, I.M. (1989) 'Throwing like a girl', in J. Allen and I.M. Young (eds) *The Thinking Muse*, Bloomington: Indiana University Press.

Young, K. (1993) *Planning Development with Women: Making a World of Difference*, London: Macmillan.

# 4 The politics of resistance

## Working-class women in rural Taiwan

*Rita S. Gallin*

The lifting of martial law in Taiwan in 1987 was followed by the emergence of a variety of social movements, including a feminist movement committed 'to change the status quo of gender relations' on the island (Ku 1989: 12; see also Ku 1989; Lin 1994). In addition to pressing for the revision of laws that discriminate against women, members of the movement have demonstrated in the streets to protest against child prostitution, wife abuse and sexual harassment in the schools and workplace. While some feminists argue that efforts such as these have set in motion a reconfiguration of 'values about the sexes' (Lin 1994: 7), most agree that 'traditional constraints are still strong and ubiquitous [in Taiwan and] equal partnership for women and men ... remains a distant objective to work toward' (Ku 1988: 186).

Many also agree that the movement faces challenges other than obdurate male structures. For example, Lee Yuan-chen, founder of the Awakening Foundation (a women's rights group), believes that women who do not 'take seriously their own rights ... and try to get reasonable treatment' pose an obstacle to the realisation of the movement's goals (quoted in Lin 1994: 7). Ku Yen-lin (1989: 22) maintains that 'women motivated by incentives other than the feminist cause' have slowed the movement's attempts to 'subvert the status quo'. Still others insist that the homogeneity of the movement's membership – largely middle-class, educated, and urban – has frustrated efforts to eliminate the systems of domination within which Taiwan's women are embedded. In the view of feminists such as these, if the movement is to successfully change the structure of gender relations in Taiwan, it must reach, mobilise, and 'generate "correct consciousness"' among 'working-class women' (Ku 1989: 22).

Part of the problem with this prescription is that it is based on the presumption that working-class women lack a 'feminist' consciousness and that political struggle occurs solely within the framework of 'organised' movements. It fails to recognise that resistance accompanies all forms of domination and that politics permeates everyday life. Moreover, this prescription overlooks the fact that although 'minor forms of resistance' may not lead to important structural transformations, they can produce changes that 'have revolutionary implications for how people lead their daily lives and construct and reconfigure their worlds' (Alvarez and Escobar 1992: 327). The purpose of this chapter is to discuss one such minor form of resistance – the practice of politics by working-class women in rural Taiwan.[1]

The chapter is based on data collected in Hsin Hsing, a small village in Chang-hua County which has changed over the past thirty years from an economic system primarily based on agriculture to one predominantly dependent on off-farm employment. The data were primarily collected during an eight-month field study in 1989–90, although I draw on insights acquired in six separate field trips to the village between 1957 and 1982. The data are based on both anthropological and sociological techniques and include participant observation, in-depth interviews, surveys and collection of official statistics contained in official family, land and economic records.

I begin the chapter by describing Taiwan to establish the economic and ideological context within which the women are embedded. Then I discuss production, reproduction, and family in the village and examine the way married women deal with, manipulate and resist power in their everyday lives. In the conclusion, I interrogate the notion of power and argue that working-class women in Taiwan should not be viewed solely as the objects of feminist practice. Rather, they should be considered subjects who are actively engaged in feminist projects.

## Taiwan

When the Nationalist Government retreated to Taiwan in 1949, it found the island to be primarily agricultural with conditions not consistently favourable to development. The strategies it adopted to foster economic growth have been documented in detail elsewhere (Ho 1978; Wu 1990). Here it need only be emphasised that the government initially strengthened agriculture to provide a base for industrialisation, pursued a strategy of import substitution for a brief period during the 1950s, and then in the 1960s adopted a policy of industrialisation through export.

The latter policy produced dramatic changes in Taiwan's economic structure. The contribution of agriculture to the net domestic product declined from 36 per cent in 1952 to only 7 per cent in 1986, while that of industry rose from 18 to 47 per cent during the same period. Trade expanded greatly, increasing in value from US$303 million in 1952 to US$64 billion in 1986. The contribution of exports to the volume of trade also rose dramatically, from US$116 million (38 per cent) in 1952 to US$40 billion (63 per cent) in 1986 (Lu 1987: 2).[2]

Changes such as these might well be documented for other countries that adopted policies of labour-intensive, export-oriented industrialisation. In contrast to other developing countries, however, Taiwan's planners did not depend primarily on direct foreign investment to stimulate growth. Rather, they relied on capital mobilisation within the domestic private sector and an elaborate system of subcontracting to spearhead the growth of manufactured exports. In fact, subcontracting is so thoroughly institutionalised on the island that one foreign executive (as quoted in Sease 1987: 1) remarked that Taiwan 'is not an exporting nation ... [but rather] is simply a collection of international subcontractors for the American market'. This assessment notwithstanding, Taiwan's industrial structure is based on and sustained by vertically integrated and geographically dispersed small-scale businesses.

As early as 1971, for example, 50 per cent of the industrial and commercial establishments and 55 per cent of the manufacturing firms in Taiwan were located in rural areas (Ho 1979).[3] Most such businesses are small and medium-sized operations that produce for domestic and international markets; more than 90 per cent of the island's enterprises each employ fewer than thirty workers (Bello and Rosenfeld 1990: 219) and, in 1990, these small businesses employed three-quarters (75.3 per cent) of Taiwan's labour force (Directorate-General of Budget, Accounting and Statistics (DGBAS) 1991: 116–17).[4] Furthermore, these small businesses are the mainstay of the island's trading economy, accounting for 65 per cent of exports in the late 1980s (Bello and Rosenfeld 1990: 241).[5]

The predominant form these enterprises take is the family firm. In fact, 97 per cent of all businesses owned by Taiwanese are family organised (Greenhalgh 1980: 13). Moreover, their personnel tend to be divided into two classes: male entrepreneurs who organise labour, and women workers who provide labour with or without wages (Gallin 1996). This division of labour, however, is not surprising. After all, the Taiwanese family is based on hierarchical male principles.

A patrilineal patriarchy, the domestic unit in Taiwan, is deeply inscribed with Confucian morality.[6] Within this moral order, women's submission to men is one of the five major human relationships. Traditionally, this idea translated into the expectation that women would obey their fathers before marriage, their husbands after marriage, and their sons in widowhood. While Taiwan's constitution officially abrogates this expectation by guaranteeing women equality with men, women continue to be expected to embody the strengths of 'obedience ... reticence, [and] adaptability' (Lang 1946: 43, quoting Confucius) and to fulfil traditional wifely and maternal roles, even if they are highly educated (Tzou 1999).

These expectations are not simply a reflection of traditional culture. The state has actively intervened to reinforce Confucian ideology in Taiwan (Chiang and Ku 1985). Educational curricula construct women and men as 'two mutually exclusive categories' (Farris 1986: 17; see also Gates 1987: 148–50), defining femininity and masculinity as attributes of biology and central to the harmonious and efficient functioning of the family. Community development programmes, which aim to mobilise women's assumed 'unused' and 'underused' labour, advocate a feminine ethic and define women's primary role as wife and mother (Hsiung 1996: 48–55). Public discourse maintains that a proper woman, like the paragon encoded in Confucian philosophy, acquiesces to the 'shoulds' and 'oughts' considered natural to female existence and accepts an image of herself as less important than men (Tzou 1999). In sum, the government, by affirming Confucian virtues and values, creates women (and men) who constitute the type of work force required by Taiwan's political economy.

## Hsin Hsing village

Hsin Hsing is a nucleated village approximately 200 km southwest of Taipei, Taiwan's capital city, and it is located beside a road that runs between two market towns, Lukang and Ch'i-hu. Its people, like most in the area, are Hokkien (Minnan) speakers whose ancestors migrated from Fukien, China, several hundred years ago.

In 1990, 457 people, in seventy-six households, were members of the village. Approximately four-fifths (79 per cent) of the population was 48 years of age or younger, and the proportion of men (51.9 per cent) was slightly larger than that of women (48.1 per cent) (see Table 4.1). Complex rather than simple type families predominated in the village. Almost two-thirds (62 per cent) of households (66 per cent of the population) consisted of parents, one or more married sons and their wives and children. Regardless of the form they took, the majority of village families (85 per cent) derived the bulk of their livelihood from off-farm employment, although a large proportion (84 per cent) continued to farm (Gallin 1995).[7]

As might be expected, off-farm work resulted in economic differentiation. Hsin Hsing included members of the proletariat who were protected by government labour codes and received wages determined by contract, and members of the sub-proletariat who, not benefiting from government legislation, received casual rather than protected wages. The village also included members of the petty bourgeoisie and bourgeoisie; each owned the means of production but the first did not hire labour whereas the second had control over labour power. The multiple reasons for a family's location in this hierarchy defy easy generalisation, but women's and men's roles in production and reproduction were integral to its formation.

*Table 4.1*  Population of Hsin Hsing village by gender and age, 1989–90

| Age (years) | Male | | Female | | Row total | |
|---|---|---|---|---|---|---|
| | no. | % | no. | % | no. | % |
| 1–17 | 81 | 34.2 | 72 | 37.2 | 153 | 33.5 |
| 18–26 | 46 | 19.4 | 34 | 15.4 | 80 | 17.5 |
| 27–48 | 60 | 25.3 | 58 | 26.4 | 118 | 25.8 |
| 49–64 | 28 | 11.8 | 29 | 13.2 | 57 | 12.5 |
| 65 and older | 22 | 19.3 | 27 | 12.3 | 49 | 10.7 |
| Total | 237 | 100 | 220 | 100 | 457 | 100 |

*Source*  Field survey

*Note*  The following rationale was used to construct age groups: 1–17 (pre-school through middle school); 18–26 (single and working or in school); 27–48 (younger married); 49–64 (older married); 65 and older (retired).

## Production, reproduction and family in Hsin Hsing

One travels by bus from Lukang to Hsin Hsing on a cement road flanked by clusters of village houses, farmland and more than fifty factories. These labour-intensive companies produce for foreign and domestic markets: they include large establishments that manufacture mirrors and venetian blinds, medium-sized enterprises that produce furniture and clothing, and small factories and workshops that either operate as subcontractors for larger firms or produce directly for the domestic market. In addition to the factories situated along the road, the area is dotted with other establishments that produce commodities for foreign and domestic consumption. Many are located in or next to their owners' homes.

This was the case for eleven factories, two workshops and nineteen sales and services shops owned by Hsin Hsing villagers in 1990. The diversity of these enterprises can be seen in Table 4.2, as can their tendency to be controlled on the basis of gender. Fully four-fifths of village enterprises were owned by men, who also tended to operate businesses which were larger in scale and more highly capitalised than those of women. Only one woman, the producer of the tongues of sports shoes, owned a firm of a calibre comparable to men's.

Nine of the factory owners produced for export. Eight were subcontractors whose production represented a sub-process in the creation of a final product for export, while one, the toy and novelty manufacturer, produced a complete product which he exported with the help of a trading partner. Seven factory owners employed waged workers, and one was the woman who produced tongues of sports shoes. All of these employers paid their male workers monthly wages and their female workers by the piece. 'Women's hours are uncertain,' one male employer explained. 'Sometimes they arrive late and leave early to take care of housework.' Four of the factory owners also employed outworkers, supplying them with raw materials and paying them on a piece-rate basis at a level lower than the minimum wage (NT$8,000 (US$290) per month in 1989–90).

Regardless of whether or not men employed workers, most were assisted in their businesses by family members, who were predominantly their wives. Indeed, only ten (25.6 per cent) of the men operating businesses did so without wage or non-wage workers. Women entrepreneurs, in contrast, tended to manage their businesses single-handedly. Seven (87.5 per cent) of the eight women owning businesses operated small-scale enterprises which they managed by themselves.

Because village enterprises tended to be owned by men, it is not surprising that slightly over one-quarter (27.8 per cent) of the married men in the village identified themselves as entrepreneurs in the industrial and agricultural sectors (23.2 and 4.6 per cent respectively) in comparison to 6.7 per cent of the married women (see Table 4.3). Similarly, it is not surprising that married women were four times more likely than men to be workers in a family business and to report that they worked without wages. Only two (9.5 per cent) of the twenty-one women working in family businesses were paid for their efforts; both were newly

*Table 4.2*  Enterprises operated by Hsin Hsing villagers, 1989–90

| Type of enterprise | Number | Type of enterprise | Number |
|---|---|---|---|
| **Factories and workshops** | | **Sales and services** | |
| Toys and novelties[b, c] | 1 | Grocery stores[a] | 4 |
| Auto mirrors[b, c] | 1 | Barber shop[a] | 1 |
| Auto oil seals[b, c] | 1 | Beauty parlour[a] | 1 |
| Decorative pillows | 1 | Motorcycle repair and sales | 2 |
| Umbrella frames[b, c] | 1 | Tailor shop | 1 |
| Nylon athletic rope finishing[b] | 1 | Chinese medicine shop | 1 |
| Suitcase construction | 1 | Pinball parlour[a] | 1 |
| Sports shoe tongues[a, b, c] | 1 | Pesticide shop | 1 |
| Metal finishing[b, c] | 2 | Betel nut vending[a] | 3 |
| Custom iron springs[b, c] | 1 | Taxi service | 1 |
| Wire sealing[b] | 1 | Interior design and decoration[a] | 1 |
| Puffed rice candy and cereal | 1 | Fried chicken vending[a] | 1 |
| | | Juice vending | 1 |
| **Total** | **13 (27.7%)** | **Total** | **19 (40.4%)** |
| **Agriculture-related** | | **Other** | |
| Itinerant vegetable sales | 3 | Construction and masonry[c] | 1 |
| Rice mill | 1 | Gambling (numbers games) | 4 |
| Grape farms | 2 | | |
| Pig farm | 1 | | |
| Duck farm | 1 | | |
| Vegetable farm | 1 | | |
| Farm labour brokerage | 1 | | |
| **Total** | **10 (21.3%)** | **Total** | **5 (10.6%)** |
| Total enterprises | 47 | | |

*Source*  Field survey

*Notes*
a   With the exception of three grocery stores operated by men, these enterprises were operated by women.
b   The owners of these factories produced for export. With the exception of the man producing toys and novelties, all were subcontractors.
c   With the exception of one metal finisher, the owners of these enterprises hired waged labour.

*Table 4.3*  Primary occupation of married Hsin Hsing villagers by gender, 1989–90

| Primary occupation | Male | | Female | | Row total | |
|---|---|---|---|---|---|---|
| | no. | % | no. | % | no. | % |
| Wage worker | 39 | 36.1 | 57 | 47.5 | 96 | 41.9 |
| Entrepreneur | 25 | 23.2 | 8 | 6.7 | 33 | 14.8 |
| Worker in family business | 5 | 4.6 | 21 | 17.5 | 26 | 11.4 |
| Farmer/Marketer | 5 | 4.6 | | | 5 | 2.2 |
| Farmer | 27 | 25.0 | 11 | 9.2 | 38 | 16.6 |
| Soldier | 2 | 1.9 | | | 2 | 0.9 |
| Housekeeper | | | 13 | 10.8 | 13 | 5.7 |
| Retiree | 5 | 4.6 | 10 | 8.3 | 15 | 6.5 |
| Total | 108 | 100 | 120 | 100 | 228 | 100 |

*Source*  Field survey

*Note*  In addition to married villagers, the figures include five widowed men and twenty-one separated women and widows.

married women who worked for their fathers-in-law, and each received NT$3,000 (US$75) per month for her labour. Three (60 per cent) of the five men working in family businesses, in contrast, were paid for their efforts.

Women who worked without wages frequently voiced dissatisfaction with the arrangement. Some women's dissatisfaction was rooted in the knowledge that others were paid for performing similar work, and often for working fewer hours. As a 46-year-old woman reported:

> I work eight hours a day and if we're busy, I … might work until 1.00 or 2.00 in the morning. When you export, you must meet deadlines. The only difference between me and the workers is that I don't get a wage.

Other women were dissatisfied because, as the woman just quoted observed, their husbands were withholding a resource which gave a woman a degree of control over her life and a measure of self-esteem. Still other women who worked without wages, as well as women who worked for wages, resented the inequitable division of labour the arrangement imposed. 'Women have two jobs – work and housework,' a 36-year-old woman complained. 'A husband only has one job. It's not fair!'

While this woman's analysis of the gender division of labour in the village was correct, few married women identified themselves as housekeepers (see Table 4.3). In fact, only one-fifth (22.5 per cent) reported that housework was something they did.[8] Nevertheless, domestic work was women's work, and men were unwilling to accept responsibility for housework. Rather, they identified productive labour as their primary role, as a 28-year-old man confirmed:

Men aren't willing to do housework. If you asked ten men, you wouldn't find three willing to do housework. I wouldn't do it.

*Interviewer:* Why?

*Respondent:* If I stayed in the house, I would think I was inferior. Women have to do housework. It's the traditional way. It's *natural* for a man to earn money.

Domestic work was also women's work because, as this young man illuminated, villagers believed that society was constructed in ways such that it was 'natural' for women, as wives and mothers, to assume all responsibility for the maintenance of the household and family. Nevertheless, although reproduction was an integral part of the wife and mother roles, neither women nor men disassociated these roles from the work role. Working was central to the wife/mother role, and a good wife

helps her husband with whatever he needs. If a husband can't do something, then the wife has to help.

*Interviewer:* Help?

*Respondent:* Work and make money. Everyone agrees that you should make money. Women don't have to ask their husbands when they go out to work ... There's no need to discuss going to work since you work to make money. The money is for family living.

(Factory worker, 29 years of age)

Working off-farm, then, both fell within the definition of proper 'wifehood' and was necessary, legitimate and an integral part of caring for the family.

Cash was required to care for the family because the rural economy was highly commoditised. The money which entered the household in the form of husbands' earnings, however, was not always adequate to pay for the wide range of expenditures incurred monthly by village families. Women thus worked to secure a basic standard of living for the family, using their earnings to subsidise the expenses of daily life.[9] Perhaps for this reason, a majority said that they did not have 'private money' (*sai-khia*).

Some younger women, however, were able to retain their earnings because of their families' relatively stable economic condition (Gallin and Gallin 1982). This cache of funds served two important functions: it enabled women to create a minimum space for control over their lives; and it helped them to gain a measure of self-respect. These functions are illustrated well in the following excerpts from my field notes:

When a woman has private money, she can have her own opinion. She can speak louder. If she wants something, she can say, 'I'm using my own money!'

(34-year-old entrepreneur)

If you have private money, you don't have to ask your husband for money.
You can have self-esteem. You don't have to be raised – like a child – by
your husband.

(29-year-old non-wage worker in family business)

Work for wages, then, accorded some women control over money, providing
them with a means to negotiate certain male prerogatives embedded in the
conjugal relationship. Such bargaining power was extremely important because
women's attempts to change other more intimate areas of the husband/wife
relationship were not always effective. Although women held precise normative
expectations of the husband's role and about 'proper' interaction between a
husband and wife, the reality of married life was that these expectations were
not always met.

Older women, for example, focused on the 'shoulds' and 'oughts' linked to
the viability of the family when they defined a 'good husband'.[10] Their descriptions were peppered with phrases such as 'he should be responsible for his
family', 'he should take care of his wife and children', and he shouldn't 'gamble',
'drink', 'spend money carelessly', 'play around' and/or 'abuse his wife'. Yet a
considerable number made abundantly clear in their discussions of the dynamics
of family life that the deference and obedience they showed their husbands was
not necessarily reciprocated with the responsibility and consideration they
expected. Rather, their husbands many times squandered family funds playing
the lottery and visiting wine houses, and humiliated them with verbal and, in
some instances, physical abuse.

Younger women, in contrast, focused on interactional behaviours in their
descriptions of the 'good husband'. In their view, the ideal mate was a
'companion', a 'thoughtful', 'considerate' and 'concerned' man who 'communi-
cated openly' with his wife and avoided fighting and quarrelling with her. Their
narratives about the realities of their lives, however, revealed that the harmony,
solicitude and affection they desired were not always forthcoming. Rather, their
husbands' tempers were apt to erupt and a couple's relationship was often
marked by petulance on the part of the husband and compromise on the part of
the wife.

Views such as these constitute an interpretation of the world that seems a
simple mystification of Taiwan's dominant ideological system. Before we accept
this notion, however, it may be useful to recall that

a woman's discussion of her life may combine two separate, often conflicting
perspectives: one framed in concepts and values that reflect man's dominant
position in the culture, and one informed by the more immediate realities of
a woman's personal experience.

(Anderson and Jack 1991: 11)

In addition to conveying different pictures of reality, however, I maintain that
these 'two separate … conflicting perspectives' constitute a mechanism by which

women engage power.[11] That is, women use language to resist cultural under-standings and to gain control of as large a part of a situation as possible.

## Women's consciousness and discursive practices

The idea that Taiwanese women use language to attain goals is not new. Writing about the Taiwanese agricultural village she studied in the late 1950s, Margery Wolf (1972) argued that women constituted a 'community', using 'talk' to curb the abusive behaviour and unjust actions of husbands. Such a community also existed in Hsin Hsing during the same period. By 1990, however, the separate groups that made up the women's community in Hsin Hsing were no longer inclusive of all generations. Rather, they were age graded. All but one included older women. The exception was made up of thirteen women in their thirties and two in their early forties. While the older women had formed their groups on the basis of friendship, the younger women had come together under the aegis of men.

In 1989, Hsin Hsing erected an elaborate temple in which to house the statues of the village god and other deities they worshipped. Because a neigh-bouring village sponsored a group of women who chanted during communal religious rites, the governing board of the temple decided that Hsin Hsing also should have such a conglomerate. They thus hired a teacher and invited women from every family in the village to join the nascent group. While thirty women originally enrolled, by 1990 only fifteen women remained members.[12]

Despite the fact that the fifteen women had no particular ties as a group when the classes began, over time they developed a 'special friendship'. As one chanter explained,

> We often practise together and then talk after practice. Talking makes you closer. We women are very talkative. There's a saying, 'Three women can make a market.'[13]

There is also a saying, according to several older women, that 'three women produce words [while] three men produce smoke'. Put another way, when women came together in a group they talked. They aired their grievances and they offered each other solace and support. Perhaps more importantly, by letting their resentments be known they enrolled other women in the project of curbing male behaviour they found intolerable.

Margery Wolf (1972: 40) explains it this way:

> Taiwanese women can and do make use of their collective power to lose face for their menfolk ... [Men] are never free of their own concept, face. We once asked a male friend ... just what 'having face' amounted to. He replied, 'When no one is talking about a family, you can say it has face.' This is precisely where women wield their power. When a man behaves in a way that they consider wrong, they talk about him – not only among themselves,

but to their sons and husbands. No one 'tells him how to mind his own business', but it becomes abundantly clear that he is losing face and by continuing in this manner may bring shame to the family of his ancestors and descendants. Few men will risk that.

The women of Hsin Hsing continued to use this strategy to control men's behaviour, as a 69-year-old woman revealed:

Women talk ... That's still a way of controlling others. Others change ... men's behaviour.

*Interviewer:* Others?
*Respondent:* Women talk to certain people – those they are familiar with. Men ... listen to their close friends or to ... their parents and relatives.

What Wolf is arguing, then, is that individual women did not possess power, but that they negotiated power by channelling information through the women's community. This collective, by intervening in the flow of events, helped women control their husbands' behaviour, if not in the long term, at least at the time it was most objectionable.

I argue, however, that the 'talk' the women's community engendered was not simply a discursive form used by individual women to manipulate problematic situations. Such talk also politicised women, enabling them to resist the forms of power present in Taiwanese patriarchy. In my view, the meetings of the community's constituent groups constituted a form of consciousness-raising – although the women certainly never called them that and they probably never thought of them in this way.

One member of the chanting group, however, remarked that 'you need to communicate [with friends] if you want to understand'. She thus was apparently cognisant that the creation of consciousness is a process that begins when women become aware and develop an understanding of the roots of a problem. In fact, the chanters were far less likely to invoke the notion of 'fate' or an individual's 'lot in life' when discussing husband/wife dynamics than were non-members of the group. Rather, they attributed discord between husband and wife to an imbalance of power created by societal rules. As one 34-year-old chanter explained, 'In a family, one is the boss and the other is the wife of the boss ... Society made it that way.' While her remark might be interpreted as a sign of resignation, in my view it denotes understanding of the hierarchical ordering of male/female relations in a patrilineal patriarchy as well as an awareness that, in contrast to inequities created by 'fate' or the circumstances of one's life, imbalances of power between men and women were 'man-made' and thus potentially susceptible to change.

Regardless of interpretation, at their meetings women unveiled their ambivalence about their husbands, marriages and families. Their perceptions of these experiences – perceptions that were usually censured by the culture – illuminated aspects of human relations that might otherwise have not been visible. One woman's articulation of injustice brought to mind other injustices for the whole group. Individual problems became social problems. The shared knowledge created helped women develop critical perspectives on the world in which they lived. This shared knowledge politicised women, enabling them to resist power on the patriarchal terrain of daily life. The form this resistance took, however, tended to vary on the basis of age.

## Older women's discursive practice

Older women appeared emboldened by their expanded consciousness, often publicly contesting the self that culture defined for them. While they were expected to be submissive and deferential to men, they often 'answered back' and acted in ways contrary to this dominant rule. On one occasion, for example, a 68-year-old man was telling me, and the men as well as other women present, how his wife had been adopted by his family and later married to him. As he related the story, his wife yelled out in an angry voice, 'My mother arranged it. Do you think I'd be so crazy to marry him?'

On yet another occasion, a man began to regale those assembled by joking about the amount of housework he did. He was interrupted, however, by his wife, who shouted, 'Liar. I do everything and you do nothing.' When I turned to her and asked why, she stridently answered, 'Ask him!' On a third occasion, I asked a group of women and men how the villagers showed affection. While it might be expected that the women would comply with cultural rules and wait for the men to speak, in this case one woman exclaimed with alacrity, 'The husband *doesn't* beat his wife!'

But older women did more than defy cultural rules. They challenged dominant gender representations. Public discourse constructs women and men as opposites, insisting that abilities and dispositions are innate and that gender roles are a by-product of biological differences. In this view, men are inherently capable and responsible 'outside people' while women are unaccomplished and pliant 'inside people'. Although strands of this dominant perspective are woven through my field notes, they also include antithetic passages in which older women describe men as ineffectual and lazy wastrels and themselves as resolute and skilled workers who engage in paid production to support the family.

This alternative discourse, I maintain, constituted a public act of defiance by which women rejected the characterisation of themselves as the inferior 'other' and subverted the view of husbands as pillars of the family. Like the examples above, they illustrate that, when older women engaged in discourses that differed from the dominant one, they challenged prevailing meanings and constructed possibilities for alternative discourse and practice.

Most of the negotiation of identity took place in the bounds of 'womanly' spaces. For example, the temple was a gathering space for women to share their disgruntlements and to check their husbands' publicly endorsed influence over them. Gossip and other transcripts in the private space of the home, as well as in public spaces such as the temple, became the means by which rural women attempted to reconfigure values and change the status quo, albeit not as a formal activist move.

## Younger women's discursive practice

Younger women, in contrast, were rarely as unrestrained in public. Choosing to avoid confrontational politics, they employed strategies of negotiation and compromise in dealing with their husbands. This approach reflected the relatively perdurable nature of marriage in rural Taiwan. Women and men there seldom voluntarily dissolve their marriages, particularly after they reach 50 years of age (Lee *et al.* 1994: 254). One reason villagers are reluctant to end their marriages was offered by a 31-year-old chanter, who maintained that

> No one is perfect … After marriage, everything is settled. Therefore, you must bear the other person's defects … During the courtship, women won't tolerate a man's faults. After marriage, you have to stay together … Marriage is not an affair of two people only. Marriage is between two families. Therefore, the couple has to communicate and try.

In this woman's view (and that of others), marriage was inviolate because of the importance of affinal ties to a family's security and mobility (Gallin and Gallin 1985). But divorce was not a viable option for most women in Hsin Hsing for other reasons as well. First, they were unable to maintain themselves on the meagre wages they earned as off-farm workers (which ranged from less than US$100 to about US$400 a month in 1990). Second, few were willing to risk losing their children who, according to family law, belong to a husband rather than to his wife upon the dissolution of a marriage.

Consequently, to preserve the conjugal bond, younger women rarely adopted confrontational stances in their dealings with their husbands. Rather, they resisted in ways that were reconciliatory. While an older woman might respond to her husband's ill temper with commensurate fury, a younger woman tended to accommodate it, using reasoning and persuasion to defuse his anger – as the following excerpt from my field notes illustrates:

> A good wife should be considerate of her husband. If something is wrong, she should not get angry. She should explain to her husband, for example, 'We have a big family and it's not good to be angry. We must work together.' Above all [she continued], don't get angry. Keep smiling!
>
> (31-year-old chanter)

Moreover, younger women rarely contested either their husbands' authority or the meaning of 'wifehood' in public. Rather, they crafted solutions that came to terms with conventional gender divisions. For example, they publicly deferred to presumed masculine authority, declaring that 'in a family, one is the boss and the other is the wife of the boss'. They also avoided threatening a man's image of himself as the defender of the family, defining themselves as wives and mothers, even as they insisted that they '*must* earn money to maintain the family'.

By adopting this traditional image for themselves and embracing such language, younger women reinforced prevailing understandings of gender. Nevertheless, when they talked in the chanting group about their lives, younger women chipped away at hegemonic discourses about gender and developed critical perspectives on the world in which they lived. Their conscious choice to 'define' and to 'name' suggests that they viewed the world as one in the making. They thus yoked together acceptance with resistance, accommodating dominant ideology while simultaneously pursuing a better world for themselves than currently existed.

## Conclusion

In summary, I have shown how the women of Hsin Hsing accepted, ignored, accommodated, resisted and protested men's power – sometimes all at the same time. My discussion suggests that, in the case of women, defining power as the ability to work one's will despite opposition may be limited. This definition characterises power as an attribute that is either present or absent, and it thus assumes that people are either powerful or powerless. A paradigmatic view of hegemony, this definition fails to acknowledge that, in Foucault's words (1980: 142), 'there are no relations of power without resistances'. Those considered powerless are never completely passive victims. Nor are those labelled powerful completely in control of specific circumstances.

Rather, 'power is fluid' (Villarreal 1992: 256) and it involves struggle, negotiation and compromise. 'It is not only the *amount* of power that makes a difference' but, as Magdalena Villarreal (1992: 256) argues, 'the possibility of gaining an edge and pressing it home'. Power involves the ability to see social spaces that can be taken advantage of, and the skill to dominate as large a part of a situation as possible – whether this control is exercised 'front or back-stage' – in this study, the private space of the home in the village or the public space of the temple – for 'flickering moments or for long periods' – as evidenced by the more carefully crafted comments by younger women in Hsin Hsing as opposed to the diatribe delivered openly by older women (Villarreal 1992: 256). Power processes are complex processes, and even when individuals make concessions and achieve only part of their projects, they may consider their objectives fulfilled.

Accommodation and strategic compliance are a part of everyday interaction, and they are mechanisms by which women deal with and manipulate the constraining and enabling factors they encounter in their lives. Resistance, like power, is multiple. It can consist of overt, collective acts of opposition or it can be manifest in 'individual acts of subtle defiance' and 'muffled voices of opposition' (Villarreal 1992: 258).

Women are 'vocal' subjects, although it is sometimes difficult to see that, when they talk, they are actually actively resisting forms of oppression and dreaming of a better world. Nonetheless, when women assert knowledge that is outside the parameters of dominant understandings, they announce to society that a funda- mental problem exists with its cultural models and forms of behaviour. When women engage in oppositional discourses, they provide elements for the creation of alternative models and practices. Language matters because it both communicates and constructs cultural practices. In truth, language is praxis.

In Taiwan, working-class women's perception of their subordination may not necessarily conform to the view of highly educated women, and they may not aspire to similar solutions to their problems as those which feminists envision. Nevertheless, I maintain, their resistance constitutes a form of feminist practice. Feminism is a highly complex and diverse set of political practices rather than simply an organised movement committed to transformative politics. Feminism is, in the words of Saskia Wieringa (1994: 834), 'a discursive process, a process of producing meaning [and] of subverting representations of gender [and] of womanhood'.

## Notes

1  I describe the married women discussed in this chapter as 'working class' to highlight the fact that class is not a matter simply of money or a relation to the means of production. Class is also about culture; about ways of thinking and doing. The culture of working-class women (both rural and urban) is critically different from that of urban, middle-class, educated feminists who wish to 'generate "correct conscious- ness"' among them (Ku 1989: 22).

2  Despite the global economic downturn, Taiwan's economy remains relatively robust. Foreign trade in 1998 was valued at US$215.4 billion, with exports contributing US$110.64 billion to this volume. While both trade and exports declined during 1998, particularly to trading partners in Southeast Asia, the Ministry of Finance reported that 'both exports and imports ... [in 1998] exceeded the US$100 billion mark, indicating that Taiwan's foreign trade remained stable despite its weakened performance' (Shen 1999: 3).

3  The dispersal of industry to the countryside has been explained as a product of industry's desire to be near sources of low-cost labour and raw materials (Ho 1976). While true, the government encouraged this movement by refraining from protecting agricultural land until the goal of industrialisation had been achieved and farm productivity had declined. In November 1975, the government promulgated a law barring the use of certain agricultural land (i.e. grades 1–24) for purposes other than farming. Before this law only grades 1–12 had been so regulated.

4  The definition of size varies in the literature on Taiwan. Ho (1976: 57), for example, defines 'small' as fewer than ten workers, 'medium' as ten to ninety-nine workers, and 'large' as a hundred workers or more. In contrast, Stites (1982: 148) defines small as a hundred or fewer workers, while Gold (1986: 141, note 16) defines large as more than three hundred workers. For the purpose of this chapter, 'small' is defined as a business with fewer than thirty workers (such businesses are not protected by government regu- lations), 'medium' is defined as thirty to ninety-nine workers, and 'large' as a hundred workers or more.

5  In 1999, small and medium-sized family enterprises continued to dominate the island's manufacturing sector and to make a 'major contribution to export produc- tion' (Courtenay 1999: 6).

6  A patrilineal family group considers only male children descent group members with rights to family property. When women marry, they sever formal ties with their fathers' families and become members of their husbands' families.

7  Despite the implementation of land reform, agriculture remains an unprofitable venture in Taiwan. On the average, Hsin Hsing farmers realised less than NT$2,000 (US$50) from the rice they grew on 1 ha of land in 1989. Nevertheless, they continued to cultivate the land because: (1) it was a source of food (rice), (2) the mechanisation and chemicalisation of agriculture obviated the need for either a large or physically strong labour force, and (3) the decreased size of family farms – in 1989 the average acreage tilled per farming household was 0.63 ha – required less labour. (For Taiwan as a whole, the average acreage tilled was 0.79 ha (*Free China Journal* 1990: 31).

8  In addition to the thirteen women who identified housekeeping as their primary activity, fourteen others reported it as their second activity. To determine the occupations of villagers, we asked two questions: What do you do most of the time? What else do you do?

9  The meaning of 'basic' was considerable in 1990 and few villagers lived at a subsistence level. It took more than one pay cheque, however, to underwrite the cost of what were considered necessary expenditures. A woman's income thus was critical to the maintenance of her family's standard of living. While the need for her income might be expected to have implications for power relations in the household, I did not find this to be the case. Rather, as I argue elsewhere (Gallin 1995), control of money rather than contribution to the budget was pivotal to the configuration of male/female relations in the household.

10  I use the terms 'younger' and 'older' women in the discussion which follows. The meaning of age, of course, is socially constructed. In Hsin Hsing, most women became grandmothers by the age of 48 or 49. People who achieved this status were considered 'old'. I follow this custom, labelling women 49 years of age and over 'older' while designating those who were junior to them in years as 'younger'.

11  The discussion which follows focuses on power contestations that revolve around gender ideology and its norms of behaviour. A reviewer of this chapter suggested that the ideology and norms associated with female hierarchies might well have implications for female/female power relations and for gendered identities. While the argument has merit, discussion of it is beyond the scope of this chapter. (See Gallin 1994 for an exploration of power relations between women.)

12  The decline, according to one chanter, reflected the fact that many of the women who enrolled 'couldn't read and there was too much to memorise'. While men theoretically were also eligible to join the group, none did.

13  The allusion here is to a forum in which information is exchanged. For the first four months after they started learning how to chant, the young women met almost every other evening to practise, usually after 8.00 p.m. 'because women can't get to the temple earlier'. Although the frequency of their meetings decreased over time, they gathered together at least four times a month to practise and, when they were preparing for a special festival, every evening for two to three hours.

## References

Alvarez, S.E. and Escobar, A. (1992) 'Conclusion: theoretical and political horizons of change in contemporary Latin American social movements', in A. Escobar and S.E. Alvarez (eds) *The Making of Social Movements in Latin America*, Boulder: Westview Press.

Anderson, K. and Jack, D.C. (1991) 'Learning to listen: interview techniques and analyses', in S.B. Gluck and D. Patel (eds) *Women's Words: The Feminist Practice of Oral History*, London: Routledge.

Bello, W. and Rosenfeld, S. (1990) *Dragons in Distress: Asia's Miracle Economies in Crises*, San Francisco: The Institute for Food and Development.

Chiang, L.N. and Ku, Y.L. (1985) *Past and Current Status of Women in Taiwan*, Women's Research Program Monograph No. 1, Taipei: Population Studies Center, National Taiwan University.

Courtenay, P. (1999) 'APEC forum outlines plan to strengthen role of SMEs', *Free China Journal*, Taipei, 25 June.

Directorate-General of Budget, Accounting and Statistics (DGBAS) (1991) *Yearbook of Manpower Statistics, Taiwan Area, Republic of China*, Taipei: Executive Yuan.

Farris, C.S. (1986) *The Sociocultural Construction of Femininity in Contemporary Urban Taiwan*, Working Paper on Women and International Development No. 131, East Lansing: Women and International Development Program, Michigan State University.

Foucault, M. (1980) *Power/Knowledge: Selected Interviews and Other Writings 1972–1977*, translated by Colin Gordon, New York: Pantheon Harvester Press.

*Free China Journal* (1990) 'As Taiwan modernizes agriculture's face wrinkles', Taipei, 9 August.

Gallin, R.S. (1994) 'The intersection of class and age: mother-in-law/daughter-in-law relations in rural Taiwan', *Journal of Cross-Cultural Gerontology* 9, 2: 127–40.

——(1995) 'Engendered production in rural Taiwan: the ideological bonding of the public and private', in R.L. Blumberg, C.A. Rakowski, I. Tinker and M. Monteon (eds) *Engendering Wealth and Well-Being*, Boulder: Westview Press.

——(1996) 'State, gender, and the organization of business in rural Taiwan', in V.M. Moghadam (ed.) *Patriarchy and Development: Women's Positions at the End of the Twentieth Century*, Oxford: Clarendon Press.

Gallin, B. and Gallin, R.S. (1982) 'The Chinese joint family in changing rural Taiwan', in S.L. Greenblatt, R.W. Wilson and A.A. Wilson (eds) *Social Interaction in Chinese Society*, New York: Praeger.

——(1985) 'Matrilateral and affinal relationships in changing Chinese society', in J.C. Hsieh and Y.C. Chuang (eds) *Social Interaction in Chinese Society*, Taipei: Institute of Ethnology, Academia Sinica.

Gates, H. (1987) *Chinese Working-Class Lives: Getting By in Taiwan*, Ithaca: Cornell University Press.

Gold, T. (1986) *State and Society in the Taiwan Miracle*, New York: M.E. Sharpe.

Greenhalgh, S. (1980) 'Microsocial processes in the distribution of income', paper presented at the *Taiwan Political Economy Workshop*, 18–20 December, East Asia Institute, Columbia University, New York.

Ho, S.P.S. (1976) *The Rural Non-Farm Sector in Taiwan*, Studies in Employment and Rural Development No. 32, Washington DC: International Bank for Reconstruction and Development.

——(1978) *Economic Development of Taiwan, 1860–1970*, New Haven: Yale University Press.

——(1979) 'Decentralized industrialization and rural development: evidence from Taiwan', *Economic Development and Cultural Change* 28, 1: 77–96.

Hsiung, P.C. (1996) *Living Rooms as Factories: Class, Gender, and the Satellite Factory System in Taiwan*, Philadelphia: Temple University Press.

Ku, Y.L. (1988) 'The changing status of women in Taiwan: a conscious collective struggle toward equality', *Women's Studies International Forum* 11, 3: 179–86.

——(1989) 'The feminist movement in Taiwan, 1972–87', *Bulletin of Concerned Asian Scholars* 21, 1: 12–22.

Lang, O. (1946) *Chinese Family and Society*, New Haven: Yale University Press.

Lee, M.L., Thornton, A. and Lin, H.S. (1994) 'Trends in marital dissolution', in A. Thornton and H.S. Lin (eds) *Social Change and the Family in Taiwan*, Chicago: University of Chicago Press.

Lin, D. (1994) 'Out of the kitchen, into the economy', *Free China Journal*, Taipei, 11 November.

Lu, M.J. (1987) 'Protection of constitutional democracy government's goal', *Free China Journal*, Taipei, 5 October.

Sease, D. (1987) 'U.S. firms fuel Taiwan's trade surplus', *Asian Wall Street Journal*, New York, 8 June.

Shen, D. (1999) 'Foreign trade off 8.9 per cent in 1998', *Free China Journal*, Taipei, 15 January.

Stites, R. (1982) 'Small-scale industry in Yingge, Taiwan', *Modern China* 8: 147–79.

Tzou, J. (1999) 'First blueprint on policy for women', *Free China Journal*, Taipei, 17 June.

Villarreal, M. (1992) 'The poverty of practice: power, gender and intervention from an actor-oriented perspective', in N. Long and A. Long (eds) *Battlefields of Knowledge*, London: Routledge.

Wieringa, S. (1994) 'Women's interests and empowerment: gender planning reconsidered', *Development and Change* 25, 4: 829–48.

Wolf, M. (1972) *Women and the Family in Rural Taiwan*, Stanford: Stanford University Press.

Wu, R.I. (1990) 'The distinctive features of Taiwan's development', in P.L. Berger and H.H.M. Hsiao (eds) *Search of an East Asian Development Model*, New Brunswick: Transactions Publishers.

# 5 Negotiating land and livelihood

## Agency and identities in Indonesia's transmigration programme

*Rebecca Elmhirst*

## Introduction

Recent studies of gender politics in the Asia-Pacific region have indicated the limitations of viewing gender identity as a space from which women act in clear and easily specified ways (Butler 1990; Ong and Peletz 1995; Sen and Stivens 1998). Gender politics are produced, negotiated and re-visioned through a complex web of power relations out of which new subjectivities are hewn, drawing on representations from the past as well as the globalised present, that concern communal, national and class identities. This complexity is especially visible when analysis is directed to the politics of locality: to settings where, on a day-to-day basis, women are embroiled with the local implications of wider processes of change and where their actions (and passivity) are guided in complex and often contradictory ways by their multiple positionalities as mothers and daughters, farmers and factory workers, wives and citizens, and so on. This chapter draws upon feminist theorisations of agency and identity to examine the intricacies of gender politics in one particular setting: the Indonesian government's transmigration resettlement programme in North Lampung, Sumatra. As a space in which two cultures – migrants and 'indigenous'[1] people – confront one another in the context of Indonesian nation-building, Indonesia's transmigration programme provides a lens through which to view some of the ways in which gender politics are played out in one particular Asia-Pacific locality. The study reveals how gender politics are embedded in cultural politics in ways which negate the possibility of female agency and compromise the position of women in transmigration areas as gender issues become subordinate to issues of identity for both transmigrants and local people.

Transmigration involves the resettlement of land-poor Javanese migrants into less populated Outer Island areas where they endeavour to forge a livelihood (with some state aid) alongside the original inhabitants of receiving areas. The programme has attracted considerable attention for its attempts to realise a particular vision of unity, assimilation and national security that has underscored New Order (and to date, post-New Order) Indonesia as it seeks to establish cultural, economic and political uniformity across disparate 'ethnic' groups (Guinness 1994; Tirtosudarmo 1995). In general terms, as an arena for struggles over the production of national citizenship across a fragmentary archipelago, transmigration provides a focus for examining issues of identity, nation and gender politics in the context of profound political, economic and environmental change. More specifically, my purpose here is to consider why transmigration appears to have yielded so little in the way of organised collective gender-based resistance, given its coercive nature and the often negative material impact it has upon the lives of women from both the migrant and local ('indigenous') Lampung communities.

The communities that provide a setting for this study are located in the province of Lampung, on Sumatra's southern tip. The province has long been a destination for migrants from Indonesia's Inner Islands of Java, Bali and Madura. The cultural ecologies of these peoples combine with the landscapes of 'indigenous' Lampungese groups who have now become a minority in the province. Since 1980, through *Translok*, the local transmigration programme, the geography of North Lampung has transformed from what was once a sparsely settled area comprising isolated communities of shifting cultivators to settled rice cultivation by Javanese migrants and the rapid growth of large-scale agro-industrial plantations which have sprung up in tandem with the resettlement programme. The research area itself is one of the poorest in Lampung (and indeed in Indonesia) which is characterised by poor land, frequent droughts and floods, and poor communications.

The study is based on two separate periods of fieldwork (1994–5 and 1998) in the Lampungese community of Tiuh Indah and the transmigration settlement of Negara Anyar.[2] Research involved qualitative and quantitative interviews in the two villages. Livelihood surveys concerning income and welfare issues were conducted on forty transmigrant households and forty Lampungese households. In addition, qualitative interviews with male and female members of both communities (forty Lampungese women, thirty-five Lampungese men, forty Javanese women and thirty-six Javanese men) and interviews with government personnel in the locality and in Jakarta were conducted. Data were also drawn from a series of eight group discussions held with different age and gender cohorts from each community (men, women, adults and young people in groups of ten people) around experiences concerning livelihood, aspirations and identity issues in the context of the transmigration programme. While a number of themes emerged in the data, what was startlingly clear was how Javanese migrants and Lampungese 'indigenous' people are involved in a panoply of overt and covert contestations over livelihood and identity in a setting marked by changing and often difficult economic and environmental circumstances. These, and their implications for gender politics, provide the focus for what follows.

## Gender divisions in transmigration and the limits of female agency

As with other similarly interventionist rural development policies, Indonesia's transmigration programme has very different material implications for men and for women. Transmigration offers a particularly distilled version of the

Indonesian government's development effort and, as such, extends prevailing notions of gender roles in Indonesian development (Dawson 1994a). Particular ways of being a 'man' or a 'woman' are valued or rejected in the discourses and practices associated with transmigration, and by extension, in Lampung's *Translok* programme. State discourses, or what Djajadiningrat-Nieuwenhuis (1992) refers to as *ibuism*, tend to represent women as housewives or mothers of development, ascribing to them a particular role in development which is firmly located in the domestic realm in a process akin to what Mies (1986) has referred to as 'housewifisation'. Such ideas have emerged from discourses and practices in various spheres, including those associated with past ideologies of the Javanese elite (Djajadiningrat-Nieuwenhuis 1992), the historical influence of Islam (Raliby 1985) and ideas that accompanied the nationalist movement and the violent emergence of the New Order (Wieringa 1988). All of these have exalted female identities based on wifehood and motherhood. More recently, the gender-blind assumptions of western-influenced economic development have also been located as a source of some of the ideas about gender apparent in state representations (Dawson 1994a).

State *ibuism* is apparent in the ways in which women are targeted in the transmigration programme in so far as this directly affects migrant women and indirectly affects 'indigenous' women living in resettlement areas. Through the programme and associated development initiatives, a line of difference is inscribed between female homemaker and male breadwinner. While these representations concerning female identities are frequently at odds with the realities of women's lives, they nevertheless are influential in reinforcing notions of respectability and in shaping aspirations that hinge on women's domesticity. Specifically, such ideas are inherent in most of the statistical and economic concepts on which transmigration policy planning is based (Dawson 1994a). Of particular importance is the concept of the household as deployed in planning and, from this, the property relationships that are set in place. Transmigration households are generally conceived of as nuclear, headed by a male breadwinner who is identified as the decision-maker and the individual to whom land title falls and to whom assistance is directed. Women are regarded as dependants with responsibility for domestic and reproductive tasks, but without independent title to land or access to credit. Where women do hold land, it is because they are widowed or divorced, and usually the certification remains in their husband's name. Thus, in effect, the practices of land control fostered by the government in the implementation of *Translok* have served to entrench and reinforce the idea of women's dependence on men and have reproduced an image of women as housewives. While the influence of the state through its development programmes has been less keenly felt in the Lampungese community, state-level gender representations remain influential in inscribing women's role as dependent housewives and men's role as breadwinners, and the semantics of these definitions are reproduced in everyday conversation.

These prevailing ideas about appropriate roles for gendered actors in transmigration have powerful material implications for the situation of women affected by the programme in North Lampung. Widows and divorcees in both the 'indige-nous' and transmigrant communities form the poorest cohort in terms of income and property ownership. One transmigrant woman described how she had learned to 'hoe like a man' in order to feed her family, but her shame in failing to conform to a feminine ideal meant she did so at night so neighbours could not see (Elmhirst 1998). Similarly, both Lampungese and Javanese women without husbands and without title to land did not qualify for credit programmes and were often excluded from the informal networks that guarantee access to other types of assistance. In addition, transmigration has undermined many activities that were hitherto the domain of Lampungese women. For example, shifting cultivation and the gath-ering and processing of forest products are no longer possible since much of the forest is gone and land has been removed from communal management since the *Translok* programme started (Elmhirst 1997). These observations strike a chord with the gender impacts of land settlement schemes elsewhere in the world where women tend to be marginalised from production since many of their former entitlements (to land, labour or livelihood) are displaced in resettlement planning, or, by contrast, they face extra work because of a labour shortage in remote and difficult areas (Schrijvers 1988; Townsend *et al.* 1995). Women may also find that access to independent non-farm incomes has been curtailed, either deliberately on the part of settlement planners (for example, in situations where marketing is tightly controlled by government authorities) or through isolation. Both Lampungese villagers and Javanese transmigrants have felt the material impact of the transmigration programme in ways that are gender-specific.

Yet despite the particular material implications of transmigration for women in North Lampung (and in Indonesia more generally) there is little evidence of women taking action *as women* in order to challenge the negative gender impacts of the programme. Where resistance does emerge, it is generally in the form of individual acts played out in the household between wives and their husbands, rather than against other actors within the transmigration programme (Dawson 1994b). It is tempting to assume that this lack of action by women as women is evidence of the ways in which Indonesian women in general (and particularly poor women from outside Indonesia's urbanised middle class) are depoliticised. But that is to ignore a long history of gender politics in Indonesia that challenges the notion that women have been silenced by traditional gender roles, by poverty or by political authoritarianism. Indonesian women are far from silent, as demon-strated by works such as Wieringa's (1988) on Indonesia's feminist movement and Saptari's (1995) work on labour activism. Indeed, casual observation of the various activisms that have emerged out of Indonesia's recent economic and political turmoil have pointed to the importance of female agency as people artic-ulate their demands *as women*.[3] There is evidence that gender identity can provide a vehicle for women's action even if women do not act as a cohesive community or in anticipated ways. Rather than focusing on the ways in which women might be depoliticised as a group, it is more useful to consider the specific character of the politics of locality in transmigration areas that lessens the likelihood of gender identity being a space for the political mobilisation of women.

## Exploring agency, identity and resistance

One way of accounting for why women appear not to be articulating their gender interests in clear and easily specified ways is to rethink the links between subject position and human agency. The adoption of more nuanced concepts of human agency marks an attempt to avoid presenting people simply as the bearers of discrete class, gender or ethnic subject positions from whence particular actions (resistance or accommodation, defiance or passivity) can be predicted (Butler 1990). Rather than highlight particular bases for action from which various acts follow, a more dynamic conceptualisation of subjectivity and agency is required that illustrates how people occupy a range of subject positions and hold a range of social interests which are fluid and, at times, contradictory (Hart 1991). Two points are important here. First, feminist and postcolonial critiques have pointed to the intersectional nature of subject positions as gender identities intersect with other sources of identity in complex ways (Mohanty 1991; Barrett and Phillips 1992). Lived experience generally exceeds class, gender or 'ethnic' categories as the meaning of these and the experiences and goals of women are variously shaped by ideologies of religion, development and nationhood in particular ways at particular times and places. In rural North Lampung, gender identities are intersectional, shaped by global, regional, national, local and individual representations. For example, discourses of development associated with Indonesia's status as a newly emerging growth economy (or, more recently, as an arena of crisis) and those associated with nation-building articulate particular gendered notions of personhood and personal conduct in ways that find material expression through the policies and practices of the transmigration programme. Similarly, gender identities emerge within and through discourses associated with regional formations such as a reinvigorated Islam or through the idea of an emergent Southeast Asia economy, or through village-level discourses around social moralities and obligations, each of these expressing particular versions of femininity. At all of these levels, subjectivities are produced through struggles around the discourses associated with work (divisions of labour and ideas about who can and cannot do what, and the meanings ascribed to particular activities), family life (including normative codes concerning gender and generational rights and obligations) and neighbourhood life (values and moral obligations).

Second, recent approaches to agency and subjectivity have pointed to the ways in which such intersectional subjectivities are 'tenuously constituted in time' (Butler 1990: 140). In other words, particular spaces of identification such as age, gender or cultural identification become significant as occasion demands and are thus contextual. This is important, as it points to how gender identity as a space from which women act as women may be subsumed as other forms of identification take precedence to meet particular political ends. Indeed, individuals are not simple products of power but are also its agents, exercising power in producing themselves as subjects, as well as producing social collectivities around different forms of identity (Escobar 1995; Sangren 1995). In transmigration, women may define themselves or find themselves defined simultaneously as a transmigrant/original inhabitant (*orang trans/orang asli*), along the lines suggested by kinship structures (mother, daughter, sister, wife), or via definitions wrought in organised religion (as Muslim or Christian). Colonial categories also remain important as people are defined according to ethnic categories based on the Dutch colonial administration's interpretation of 'indigenous' customary law in the nineteenth century.[4] Other identities emerge through livelihood discourses in which people define themselves or are defined: for example, as farmers (*petani*), labourers (*buruh*), business people (*orang bisnis*) or, following contemporary discourses of femininity, housewives (*ibu rumah tangga*). However, it is the particularities of the *Translok* programme in North Lampung that disturb a simple interpretation of female agency as gender identities are muddied by a more complex cultural politics.

## The cultural politics of transmigration

One of the paradoxes of Indonesia's transmigration programme, and by extension *Translok* in North Lampung, is that while it is aimed at unifying a diverse and disparate archipelago, the programme created and reinforced a divide between Javanese migrants and, in this case, the Lampungese local people. Contrary to the cultural, economic and environmental uniformity that transmigration is supposed to bring in its wake, resettlement areas are spaces in which the tenor of cultural politics is raised: on the one hand, because of inequalities bound up in the way the programme is administered, and on the other, as 'border-zone' politics are played out between Lampungese and Javanese people themselves as they engage in conflicts over resources, meanings and representations.

### The politics and practices of transmigration

Within *Translok* administration, Lampungese local people and Javanese migrants are treated rather differently, both discursively and in a material sense. First, as a means of providing land-poor Javanese families with their own plots of land, *Translok* enables transmigrants to obtain land to which they hold official certified title. By contrast, Lampungese people have seen their former community land nationalised while they have yet to receive official title to their privately owned plots.[5] Effectively, this has excluded Lampungese people from resources that are only accessible with official land title (such as access to credit) and has set in train ongoing disagreements over land ownership between government, Javanese migrants and indigenous Lampungese people. Cultural differences are thus underscored and reinforced through *Translok* resource politics.

Second, *Translok* is a centrepiece of government social engineering, in Lampung as elsewhere in Indonesia, involving the delivery of various agricultural development programmes and new village administrative structures that extend a particular vision of modernity and development across the archipelago (Guinness 1994). While it is in transmigration settlements themselves that this model is realised, the processes and practices associated with it are expected to filter across into surrounding areas, transforming them and bringing them into line with the state's vision of Indonesian modernisation (Dove 1985). In practice, such efforts seal in a line of difference between migrants and local people as transmigration land use and administrative models stand in marked contrast to local practices. Although such models are expected to spread into neighbouring communities, in practice this is not the case. Instead, resources continue to be concentrated in transmigration settlements, again entrenching cultural differences and fuelling animosity between migrants and local people.

Finally, and related to the above, *Translok* is implicitly concerned with the security of regions regarded as 'unstable' as the programme marks an effort to produce a particular vision of national citizenship. This vision embraces ideas about appropriate roles for different gendered actors within Indonesian development discourse and practice as a whole, effectively producing 'Indonesian' subjects, and disavowing more localised ('indigenous') ethnic subjectivities (Clauss and Evers 1990; Tsing 1993; Fachry 1997). For example, particular ways of being a farmer are 'valued' by the state, which rewards those undertaking sedentary cultivation of food crops, particularly rice and the cultivation of certain plantation crops (oil palm, sugar cane), while simultaneously representing other kinds of farming (such as shifting cultivation) that are associated with non-Javanese cultural groups such as the Lampungese as 'backward' (Dove 1985; Persoon 1992). Observers have noted how these representations of citizenship are envisioned through the eyes of the country's politically dominant cultural group, the Javanese, which may indeed be an anathema to non-Javanese cultural groups (Guinness 1994; Pemberton 1994; Widodo 1995). These representations may have material implications, since the Lampungese are effectively disenfranchised (as centrally appointed Javanese officials take up posts in non-Javanese areas) and Lampungese livelihood and governance practices (including land access mechanisms) are marginalised within *Translok* development discourse as a whole. Again, a line of difference is sealed through the practices of the *Translok* programme, and it is this, rather than gender, that defines the texture of local politics for women in North Lampung.

### Difference and 'border zone' politics in transmigration

The divisions and inequalities inherent in the programme and the resentments they have provoked have ensured that *Translok* creates a micro-political context in which cultural difference matters. Within a transmigration 'border zone' (Douglas 1966; de Lauretis 1987), cultural identity becomes a powerful idiom through which struggles around livelihood and resources are articulated. Conflicts have emerged around cultural assimilation (Tirtosudarmo 1995) and over the distribution of benefits in the transmigration programme where migrants are seen to gain more than local people. There is also conflict over the imposition of particular modes of livelihood upon shifting cultivators. In North Lampung, cultural identities have become important political spaces, as Javanese migrants and Lampungese local people challenge and resist efforts to create uniformity. One way in which this is achieved is by each group highlighting their 'distance' from a generalised Lampungese or Javanese 'other'. By calling upon their contrasting 'traditions' and local moralities they become differentiated. Pressures on livelihood and resources, coupled with inequalities related to the operation of the transmigration programme, mean that what might otherwise be a celebration of diversity sometimes manifests itself as a metaphor of moral disapproval about the practices of the other group. This is expressed in the ways that Javanese transmigrants and local Lampungese people also challenge and contest one another around issues concerning livelihood and resource claims. In addition to the material aspects of transmigration cultural politics are psycho-social concerns about purity and communal boundaries that emerge around fears of the 'other': a theme that is common in other parts of the world where groups have been brought together in difficult material circumstances (Kristeva 1991; Rosario 1994; Malkki 1995).

While struggles over resources between the two groups may take an overt material form (i.e. violent conflict between migrants and Lampungese people over land) or a covert form (theft, arson and vandalism of crops), by far the most common way that the two groups interact and challenge each other's legitimacy is through the hidden (and sometimes not so hidden) 'transcripts' (Scott 1990), gossip and rumour through which one group represents and resists the actions and representations of the other. The transcripts that circulate in each community have an important effect on how people are able to conduct their lives under the eyes of neighbours watching for any sign of transgression – evidence that cultural purity has been defiled or that communal boundaries have been breached. This scrutiny of practices reinforces behaviour in line with what is deemed to be the proper conduct of individual and collective lives (Malkki 1995), especially as fear of ridicule or malicious gossip from neighbours is particularly acute in this context where extra-household relations are important for maintaining the flow of life. Central to this are concerns over appropriate *gendered* behaviours which circumscribe the boundaries of women's lives and, indeed, female agency itself.

Many of the transcripts concerning what is held to be appropriate gendered behaviour emerge around discussions of livelihood issues and livelihood practices. They thus have important material implications for men and women: a direct impact on divisions of labour in line with the different meanings associated with particular types of work, the different spaces in which work is conducted, and the kinds of behaviours associated with particular types of work. Two moral maps are commonly invoked in each community as the (imagined) practices of the other are commented on and assessed. The first of these concerns assessments as to whether behaviours exhibit reason or rationality (*akal*) which is a positive ascription, and passion or desire (*nafsu*) which is generally negative. Both terms are originally Arabic terms, suggesting a historical Islamic influence in abstract thinking about human nature, and both emerge in many cultures of the Malay–Indonesian archipelago (Siegal 1969; Peletz 1995).

Maintaining self-control (*akal*) and containing desire (*nafsu*) are both regarded as important in avoiding ridicule or losing prestige (*gengsi*). Signs that either group is more prone to allow 'reason' to be dominated by 'passion' are looked for: for example, whether the Javanese (or Lampungese) are diligent in fasting at Ramadan and whether they observe food prohibitions. As Peletz (1995: 90) points out, 'exercise of restraint is an important ethnic marker as well as an index of virtue (or its absence)'.

A second moral map that is important in the context of wider changes in Indonesia (and beyond) concerns evaluations of behaviour according to whether it is deemed traditional or modern. This particular code owes much to state-level discourses about modernity in Indonesia where the modern practices of urban centres are contrasted with the traditional ways of regional minorities. Activities and behaviours considered to be 'modern' have a positive ascription while those regarded as 'traditional' are negative. Between Javanese migrants and Lampungese people, phrases such as '*masih di belakan* (still behind)' are contrasted with '*sudah maju* (already progressive, modern)' to describe different ways of life.

Much of the scrutiny of either group concerns their different livelihood practices, such as the ways in which land is utilised, and activities deemed appropriate for men and women in terms of their accordance with the binary pairs of 'traditional/modern', and 'reason/desire (*akal/nafsu*)'. Among each group, certain types of work are regarded as traditional or modern and are evaluated as such. Similarly (and by contrast), particular types of work are considered to promote 'reason' or 'desire', either through the nature of the activity itself or through the space in which it is conducted. For example, in terms of the rice farming system, Javanese migrants place value on hard work and their association with 'modern agriculture' through Green Revolution packages, the use of which assures their status as 'progressive' (i.e. not traditional) farmers. By contrast, Lampungese rice farming, with its emphasis on traditional varieties and 'indigenous' knowledge, is considered by transmigrants to be backward and culturally anachronistic. The hard work of Javanese farmers is considered by Lampungese people to be base and coarse as field work in general is considered to be 'animal', the triumph of 'desire' over 'reason'. This representation is illustrated by the taboos in the Lampungese community that prevent certain 'vulnerable' groups from working in the fields and, by extension, in the nearby sugar plantation. Thus, while Javanese migrants supplement their incomes with plantation work, even the poorest Lampungese farmers will not pursue this option given its historical associations and construction as a base, sexualised arena in which desire is unconstrained. Instead, Lampungese people work in trade, a practice which is in turn regarded by Javanese migrants as unrefined, uncivilised and low-status: 'They [the Lampungese] are rough, all they think about is money. For myself, religion is more important,' said one male transmigrant, aged 45.

A strong sense of there being 'some things that we do, some things that they do' pervades livelihood discourses in both groups and maintains a boundary between Javanese migrants and Lampungese people as each community distances itself from the other. The ways these differences are maintained is rarely through direct confrontation, for the two groups meet infrequently. Instead, the ways in which the other group, and particularly women, is 'imagined' to behave is used as a measure against which the practices of neighbours are evaluated and scrutinised through gossip and innuendo. In this way, the border between the two communities is policed and reinforced in response to a heightened awareness of cultural difference wrought by the injustices of the *Translok* programme and psycho-social concerns over purity and communal boundaries. The importance of transmigration cultural politics, in terms of resource inequalities and 'boundary maintenance', complicate readings of female agency both among Javanese women migrants and among Lampungese women whose gendered subjectivities are bound up with wider cultural projects.

## Female agency and boundary politics in North Lampung's transmigration zone

Given that the role of female agency is all but invisible in studies of transmigration politics, it is tempting to assume that gender, as a mobilising centre for identities, is unimportant in a resettlement context where material and representational struggles are most likely to be played out through cultural idioms. As the discussion so far has indicated, the resource conflicts inherent in the transmigration programme mean that people are more likely to identify as 'Javanese' or 'Lampungese', rather than as 'men' or 'women' with gender-specific concerns and interests. However, this is not to say that gender politics are not important in North Lampung's local transmigration programme. Indeed, it can be suggested that gender is central to the cultural politics of *Translok*. Just as gender politics are inflected by wider cultural projects in this context, so too do gender politics inflect cultural struggles over resources and representations in the programme. While this inflection disrupts the possibilities for action based on any notion of shared gender interests between Javanese migrant and Lampungese communities, it does not disavow the role of female agency. Female agency is important in two related but potentially contradictory ways: first, in that contrasting gender discourses and women's practices are central for both communities as each refuses the assimilating and unifying aims of Indonesian transmigration; and second, as women themselves challenge the male power that is inherent in their representation as cultural 'boundary markers'.

With respect to the first of these issues, a number of studies have highlighted the significance of gender identities in contexts where the borders between particular social systems mark a space of contestation. In such settings, women may be seen as markers of identity or bearers of tradition (Rosario 1994). As Moghadam (1994: 16) writes, it is 'in the context of the intensification of religious, cultural, ethnic and national identity – itself a function of uneven development and change – that we see the politicisation of gender, the family and the position of women'. The same can be said in North Lampung's *Translok* area, where differences in the work practices of women are seen as important boundary markers between Javanese migrants and the Lampungese. Gender politics are thus inscribed in and through a politics of ethnic difference as women resist the ways in which transmigration projects a particular vision of Indonesian femininity and hold on instead to their own ideas about appropriate gendered behaviours (Elmhirst 1997). Instead, particular gendered cultural identities are forged that are imbued with ideas about ethnicity and communal boundaries. Women's behaviour is used to signify differences between those who belong to each community and those who do not (Yuval-Davis 1994). In both villages, gender roles and relations provide a lens through which differences between Javanese migrants and 'indigenous' Lampungese people can be observed and evaluated, according to the frameworks outlined above.

Discourses of marriage and divorce are particularly powerful realms through which the imagined behaviours of the two groups are compared and contrasted, and righteousness reaffirmed. For example, Javanese women regard the lot of Lampungese women to be a poor one in marriage: 'They are still very constrained [*terikat*]. I think it is because of their customs. For Javanese women, it is more advanced now, more free [*bebas*],' said one Javanese woman aged 42. The contrast she draws between tradition and modernity expresses what is valued by Javanese people and what distances them from their Lampungese neighbours. For Lampungese people, Javanese marriage practices, particularly the frequency of divorce, are an object of derision: 'No man would ever marry you if you'd been married before,' said one Lampungese woman aged 36. 'Javanese people marry two or three times, but for us, marriage is until death. Lampungese women are very faithful. Even if widowed, we don't marry again.'[6] Javanese women are considered to be *bebas* (free), but in this sense, the word carries a negative connotation of unbridled desire (*nafsu*).

Teenage girls are particularly important in this equation because of their unmarried status, which gives them the greatest potential to disrupt social boundaries by marrying someone from the other community. Fear that Lampungese daughters could marry Javanese men fosters a sense that their behaviour must be policed, in order that they do not succumb to the lack of 'restraint' (*akal*) displayed in Javanese marriage and divorce and thus bring shame on their families. The behaviour of daughters is very much regarded as a reflection of family prestige and morality in the eyes of neighbours. In communities, unmarried women are considered to be the responsibility of parents, in both economic and social terms.[7] Any transgressions of appropriate codes of behaviour for young women are therefore regarded as a threat to the standing of the family as a whole. In situations of economic hardship, on the one hand, and as tones of moral disapproval are adopted by each group to describe the livelihood practices of the other, controlling the activities of daughters, including their labour, becomes an issue of political importance. 'I always say to my daughter, be good, work hard, always be polite and don't tell lies,' said one Javanese mother aged 32. 'Don't bring shame on your family, for they have to work very hard for you.'

Work practices are of particular importance, especially the meanings attached to field and plantation work, and how engagement in them accords (or does not accord) with the maintenance of moral certitude. Among Javanese migrants, teenage girls' activities are constructed in line with a notional sharing of roles between men and women in forging livelihoods in difficult circumstances. Thus, a daughter's ascribed role within this is as an essential contributor to the household economy by taking part in farm and plantation work of some kind. Daughters are expected to comply with the expectation that they work, and are good, in order to uphold family status in the eyes of the community. Working hard and being responsible in striving to be part of a 'progressive farming family' (*tani maju*) is what is expected of Javanese daughters. For this reason it is common to see young women engaged in heavy field tasks on family farms and at the nearby sugar plantation. Unlike the Lampungese community, female engagement in plantation or field work carries few negative cultural connotations. It is an acceptable fall-back strategy that accords with the idea of daughters helping their families rise up out of poverty and into the realm of progressive, modern, New Order farmers. That plantation daughters are regarded as *bebas* (free) in engaging in such work is granted a positive connotation to indicate that as modern women, Javanese transmigrant teenagers can go where they like, unlike their constrained (*terikat*) Lampungese peers.

Among Lampungese people the moral map of 'reason' and 'passion' informs local gender discourses and provides guiding principles around which family and individual respectability, status and prestige are negotiated. Of particular importance is the possibility of young women, regarded as a vulnerable group in terms of moral fortitude, being unable to show 'restraint' or 'reason' or exercise self-control in particular circumstances. This idea is expressed in the tradition of female exclusion in the area and the gendering of space more generally. Young girls who have reached puberty are confined to the house, precluded from taking part in any activity that might compromise their 'reason' until they are married and with a child. Even older women are reluctant to go to the fields alone, and they would never go to work on the plantation. Fields and the plantations are considered sexualised spaces in which the possibility of giving way to desire (*nafsu*) is raised. Without the guidance of 'rational' male relatives in such spaces, women leave themselves open to stigmatisation. Even among the poorest Lampungese households, young women do not engage in plantation work to augment farm livelihoods as maintaining status through the protection of daughters is paramount: 'If you can live without going to the fields, then it is better not to,' said one Lampungese woman aged 19. 'We are ashamed to do so. With Javanese girls, it is different. They work hard in the fields. They are more free [*bebas*].' In this instance, *bebas* implies licentious behaviour. Lampungese parents report, with horror, tales of flirtations between Javanese teenagers (male and female) who crowd together on the trucks that take them to work on the plantation. The idea that their own daughters might be tempted into such behaviour that would violate their 'honour' is unthinkable. These concerns, and the imagined transgressions being made by Javanese transmigrant women, ensure that gender boundaries are policed strictly and the participation of unmarried Lampungese women in the farming system is rare. Bu Aiysah, a Lampungese woman now in her thirties, explained how seclusion had carried on into her married life:

> When I got married it was three years before I was allowed out [effectively, after her first child was born]. If I wanted to see out, it was through the door or the window. It was very boring. Just washing, cleaning, cooking, looking after the child. I never saw anyone all day.

Through differences in marriage traditions and the work practices of young women, contrasts between the two communities are thus upheld in ways that discourage any identification of shared gender interests.

### Refusing border-zone representations

While competing subjectivities partly explain the limits of gender interests as a basis for resisting transmigration, female agency in transmigration is played out in other ways as women themselves challenge the male power inherent in their representation as 'ethnomarkers'. Particularly important at the familial level are the continual challenges made by young women to male and parental power that is invoked to define appropriate behaviours for Javanese or Lampungese girls. For young women in both communities, ideas and concerns about gendered behaviour associated with different work practices weigh heavily through the assertion of parental power within the family. Parental authority is articulated through refusal to grant permission for their daughters to participate in particular activities and through the withholding of the resources (money, transportation or a network of contacts) that would enable them to take up different forms of work. While parental power has considerable efficacy, young women in both communities are nevertheless able to negotiate and subvert parental power and aspirations. This is shown most clearly in the debates and negotiations surrounding their participation as migrant labourers in factory work, a form of employment that seems antithetical to the kinds of activities and behaviours deemed appropriate within the codes and conventions that have hitherto marked the cultural boundaries between Javanese migrants and the Lampungese community.

Since about 1992, increasing numbers of young Lampungese women have left their village temporarily, taking up work in the textile factories of Greater Jakarta, suggesting that the practices of female seclusion and economic inactivity are being negotiated. Although the practice of labour migration to distant factories is much less prevalent among Javanese transmigrants, more recently several young women have made a similar move, again suggesting that ideas about daughters' commitment to the family livelihood through their work on the family farm is also under negotiation. In both communities, the initiative to migrate has come primarily from young women themselves, often in the face of parental disapproval, concern over their moral welfare and heightened fears about the gossip that factory work might prompt. Ibu Amatyani, now married, worked in a knitting factory for six months, sewing the sleeves on to sweaters for export. She went to Tangerang because she was bored hanging around at home with nothing to do. All her friends had already gone to work there, and she did not want to be left out. 'It was good in Tangerang, very busy. Here, people are very quiet, it is boring. In Tangerang there are lots of young women who are working together, and it is more fun.' Much of the money she earned, she said, was used up in coming home for the Muslim Lebaran festival and in buying presents for her family. From a long-term financial perspective, she saved nothing and gained little from the experience.

Both Javanese migrant and Lampungese parents have been reluctant to concede to their daughters going. Ibu Mastura, a Lampungese woman aged 40, explained that both she and her husband had not wanted their 19-year-old daughter to go to Tangerang: 'I am afraid she will marry someone there. I want her to come home and not go back there again. But I expect she won't want to come back.' The concerns of parents are not surprising, given the nature of factory work. The factories are about eighteen hours' drive from North Lampung, and are situated on Jakarta's industrial fringes. Tangerang, the principal area to which young women go, is made up of many factory establishments in which people from many different parts of Indonesia are employed. Most women work in textile and sewing factories in which the work is hard and the hours are long, even if they are indoors and thus protected from the sun. Most share rented rooms in boarding houses and walk to and from work together, a practice that brings them into contact with young men. Outside work hours their time is their own, and for most young women social activities involve meeting friends to go window-shopping or going to small cafés to eat. For some young women, Tangerang offers them the opportunity to meet new people, to widen their experiences and even to choose a boyfriend (possibly from another ethnic group) without seeking their parents' approval. As would be expected, among Lampungese parents there are fears about daughters' morality being tainted in the rather ambiguous space of the factory zone and about their marrying someone from another group while living at such a distance from home. Among Javanese transmigrant parents, there are concerns about losing control of their daughters over such a distance and, importantly, losing control of their daughter's income. In both instances, the fears that parents articulate reveal concerns that the conceptual distance between the two communities may be compromised as girls from both engage in a form of work that is ambiguous in terms of what it represents.

The kinds of negotiations that have taken place have involved a number of challenges to parental authority on behalf of young women from both communities. First, some young women have sold personal possessions or borrowed money from friends in order to raise the fare to get to Tangerang in the first place. Second, daughters from both communities, but particularly Lampungese daughters, have been able to make use of peer group networks in order to press their case to their parents and in order to satisfy parental concerns that they were being 'supervised' on the journey to the city and in their lives once they arrived. Finally, and related to both the preceding points, factory migration has brought a new-found confidence to young women to challenge parental authority, and indeed the gender structures of the community, in a very direct way. The leverage used by daughters to persuade their parents to go along with their plans is not so much an economic argument but one that appeals to their parents' duty to appear as modern households. Mbak Rokiah, a Lampungese woman of 18, described how she had persuaded her parents to let her go:

> They didn't have enough money to send me to senior high school even though I really wanted to continue. They knew that. I said, if I can't go to school, then I want to go to Tangerang instead. Because of that, they had to say I could go.

Lampungese daughters have been able to persuade their parents that factory work is not like agricultural work as it is 'light' and conducted indoors, nor does the space of the factory constitute a moral threat as young women work together and are supervised by older women. In the village, the imagined space of the factory is one of 'modern', 'white-collar skilled' work. Factory workers themselves are regarded as an educated, urban workforce with pale skin and soft hands, whose 'self-restraint' (*akal*) is closely supervised by surrogate moral guardians. At the same time, Javanese migrant daughters have, to some degree, shown their parents that it is possible for them to be economically self-sufficient, and in so doing to lessen the burden upon the household as a whole. Similarly, their stories have reinforced an idea of the factory as a space of modernity in which progressive and 'free' (in a positive sense) women are able to generate an income.

Although the debates and negotiations between parents and their daughters are played out rather differently in each community, common to both is a refusal on the part of young women of the cultural politics inherent in the transmigration area. In contrast to the kinds of local representations of gender expressed in *Translok*, gender identities are being reworked by young women by drawing upon an inter-regional version of 'Southeast Asian factory girl' femininity to challenge parental authority (Ong and Peletz 1995). This way of 'doing gender' (Butler 1990) involves a set of representations and behaviours that include particular notions of female beauty (wearing 'fashionable' clothes and make-up), being articulate ('I know I'm braver at talking'), being relatively worldly ('now it is me asking my daughters' opinion: imagine that!') and being relatively well-educated compared to their rural peers (most factory women are educated to junior high school level). What is intriguing, at least in terms of transmigration cultural politics, is how these representations are shared between young women from different parts of Indonesia (Lampungese or Javanese) as cultural difference appears to be less important as a space from which young women factory workers might act to challenge the material problems they face as factory workers. However, this does not mean that these young women are defining themselves in line with a pan-Indonesian version of femininity such as that extended by the Indonesian state as it attempts to reproduce permissible modes of citizenship. These young women are a long way from being the housewives (*ibu rumah tangga*) or the 'mothers of development' of Indonesian development discourse (Smyth 1993). Rather, and particularly in the wake of the economic crisis which has brought unemployment and a reinvigorated labour activism to the West Java factory zone, their gender identities may well be intersecting with class identity. For most of the young women in this study, their participation in formal labour activism had been very limited to date. However, the degree to which their gendered class identity might become a space from which they are able to articulate demands *as migrant women workers* remains to be seen.

## Conclusion

In Indonesia, as elsewhere in the world, gender identities are interwoven with a range of different positionalities which appear to condition the extent to which gender identity can be a space that motivates women's political actions. In this study and in line with recent feminist approaches to female agency that highlight the fluidity of gender identity (Butler 1990; Pratt 1993), gender intersects with other axes of identity such as ethnicity, class, generation and nationality in complex ways. An examination of the interweaving of gender and ethnicity in transmigration work practices reveals the nature of female agency in a transmigration context where women's actions arise in complex ways from fragmented and unstable gendered subject positions (Butler 1990): as migrants or 'indigenous' people, as Indonesians or as 'ethnic' subjects (Ong and Peletz 1995). It suggests that the complexities of female agency in such a setting point towards explaining why female agency is so rarely a vehicle for challenging the very real gender-specific material impacts of the transmigration programme. In a wider sense, such a study is indicative of the problems inherent in viewing gender identity as a space from which women act in clear and easily specified ways and shows how it is unhelpful to isolate gender from other hierarchised domains.

In North Lampung, where the transmigration programme has brought resource inequalities between Javanese migrants and Lampungese people, a heightened awareness of communal ('ethnic') identity has given shape to local politics, providing a focus around which people articulate their concerns. In this context, gender identity, as a political field, has tended to become buried as cultural difference divides Javanese and Lampungese women and as communal concerns are seen to take precedence over gender concerns. Yet as this study has shown, the picture is more complex: while gender concerns have been subordinated, ideas about gender roles and representations have also been crucial to the ways in which each community defines itself in relation to the 'other'. Furthermore, the dynamics of this process are revealed in the actions of young women as they challenge the parental and male power bound up in culturally specific ideas about appropriate behaviours for women, calling upon new and emerging gender representations.

The North Lampung case shows that gender identities are fluid, contested and negotiable. They are part of a complex web of power relations out of which new identities may be hewn, drawing on representations from the past (for example, female seclusion in the Lampungese community) as well as from the globalised present (for example, regionalised gender identities based on participation in factory employment). This, coupled with the limited possibilities for formal political organisation in a country beset by economic and political crises, indicates some of the problems inherent in attempting to locate 'gendered' political strategies at a local level and in identifying clearly articulated instances of female agency in Indonesia.

## Postscript

The many changes that Indonesia has undergone since the research for this chapter was completed are likely to have further complicated the arguments put forward here. Since the fall of Soeharto from the presidency there has been an upsurge in political consciousness among ordinary people, and unprecedented calls for political change, many of them involving articulate women's groups campaigning on gender issues. This, coupled with the appointment of a female vice-president (subsequently president) – Megawati Soekarnoputri – would seem to herald the demise of the 'vision of Indonesian femininity' that permeated government programmes such as transmigration under Soeharto's New Order. Indeed, some of the gendered rituals of state (for example, 'traditional' codes of dress for women at formal meetings) are beginning to give way. New possibilities for gendered identities look set to emerge. However, how these might be played out in a political climate that is marked by calls for regional autonomy (and secession) and where popular feeling in some provinces remains cool towards being governed by a Javanese president and vice-president in Jakarta, remains to be seen.

## Acknowledgements

Research for this chapter was supported by the Economic and Social Research Council (ESRC) of the UK, grant number R000222089, and by an ESRC post-graduate award. The author is grateful to the Indonesian Institute of Sciences, the Department of Forestry, Jakarta, the International Centre for Research in Agroforestry (Southeast Asia) and University of Lampung for their support in field work.

## Notes

1   'Indigenous' and 'ethnicity' are in quotation marks in recognition of the problematic nature of such essentialist terms.
2   Names have been changed to protect privacy.
3   There has been a mushrooming of women's organisations in Indonesia since overt and widespread challenges to Soeharto's rule began in the late 1990s. Most of these

have sought political changes that embrace the gender concerns of particular groups of women. For example, women organised around the issue of food prices, culminating in the 'baby milk' demonstrations in 1998 that reworked notions of *ibuism* to challenge the state's role in the monetary crisis. Those working on issues of violence against women have organised to seek justice following the widespread rape of Chinese and other minority women during recent political and economic turmoil.

4   In North Lampung, migrants are largely Javanese, while Lampungese people belong to the *Way Kanan* group, defined genealogically, territorially and by language.

5   In the early 1970s, land belonging to 'traditional' community groups that had not been brought under permanent cultivation became state land. Much of this land has subsequently been allocated for state programmes such as transmigration and plantation development. Competing claims for this land continue to be a source of conflict throughout Lampung.

6   Under customary law Lampungese women are permitted to marry the brother of their late husband.

7   Upon marriage, women become the responsibility of their husbands, and authority passes from parents to husband. Accordingly, the persistence of the practice of first marriage below the legal limit reflects a desire by impoverished parents to transfer responsibility to a son-in-law's family.

# References

Barrett, M. and Phillips, A. (1992) 'Introduction', in M. Barrett and A. Phillips (eds) *Destabilizing Theory: Contemporary Feminist Debates*, Cambridge: Polity Press.

Butler, J. (1990) *Gender Trouble: Feminism and the Subversion of Identity*, London: Routledge.

Clauss, W. and Evers, H.D. (1990) 'The social impact of transmigration: a study of a settlement area in East Kalimantan', in A.A. Saleh and D.F. von Giffen (eds) *Sociocultural Impacts of Development: Voices From the Field*, Padang, Indonesia: Andalas University Research Centre.

Dawson, G. (1994a) 'Development planning for women: the case of the Indonesian transmigration programme', *Women's Studies International Forum* 17, 1: 69–81.

——(1994b) 'Development programmes for women farmers in a transmigration settlement: gender, pragmatism and resistance', paper presented at the *Asian Studies Association of Australia Biennial Conference*, 13–16 July, Murdoch University, Australia.

de Lauretis, T. (1987) *Technologies of Gender: Essays on Theory, Film and Fiction*, Bloomington: University of Indiana Press.

Djajadiningrat-Nieuwenhuis, M. (1992) 'Ibuism and priyayization: path to power?' in E. Locher-Sholten and A. Niehof (eds) *Indonesian Women in Focus: Past and Present Notions*, Leiden: Koninklijk Instituut Voor Taal-, Land-, en Volkenkunde (KITLV) Press.

Douglas, M. (1966) *Purity and Danger: An Analysis of Concepts of Pollution and Taboo*, London: Routledge.

Dove, M. (1985) 'The agroecological mythology of the Javanese and the political economy of Indonesia', *Indonesia* 39: 1–36.

Elmhirst, R. (1997) 'Gender, environment and culture: a political ecology of transmigration in Indonesia', unpublished PhD thesis, University of London.

——(1998) 'Reconciling feminist theory and gendered resource management in Indonesia', *Area* 30, 3: 225–35.

Escobar, A. (1995) *Encountering Development: The Making and Unmaking of the Third World*, Princeton: Princeton University Press.

Fachry, A. (1997) 'Sharing a room with other nonstate cultures: the problem of Indonesian kebudayaan bernegara', in J. Schiller and B. Martin-Schiller (eds) *Imagining*

*Indonesia: Cultural Politics and Political Culture*, Monographs in International Studies, Southeast Asian Series Number 97, Athens, Ohio: Ohio University Center for International Studies.

Guinness, P. (1994) 'Local society and culture', in H. Hill (ed.) *Indonesia's New Order: The Dynamics of Socio-Economic Transformation*, Sydney: Allen and Unwin.

Hart, G. (1991) 'Engendering everyday resistance: gender, patronage and production politics in rural Malaysia', *Journal of Peasant Studies* 19: 93–121.

Kristeva, J. (1991) *Strangers to Ourselves*, New York: Columbia University Press.

Malkki, L. (1995) *Purity and Exile: Violence, Memory and National Cosmology among Hutu Refugees in Tanzania*, Princeton: Princeton University Press.

Mies, M. (1986) *Patriarchy and Accumulation on a World Scale: Women in the New International Division of Labour*, London: Zed Books.

Moghadam, V. (1994) 'Introduction. Women and identity politics in theoretical and comparative perspective', in V. Moghadam (ed.) *Identity Politics and Women: Cultural Reassertions and Feminisms in International Perspective*, Boulder: Westview Press.

Mohanty, C.T. (1991) 'Introduction: cartographies of struggle', in C.T. Mohanty, A. Russo and L. Torres (eds) *Third World Women and the Politics of Feminism*, Bloomington: Indiana University Press.

Ong, A. and Peletz, M. (eds) (1995) *Bewitching Women, Pious Men: Gender and Body Politics in Southeast Asia*, Berkeley, CA: University of California Press.

Peletz, M. (1995) 'Neither reasonable nor responsible: contrasting representations of masculinity in a Malay society', in A. Ong and M. Peletz (eds) *Bewitching Women, Pious Men: Gender and Body Politics in Southeast Asia*, Berkeley, CA: University of California Press.

Pemberton, J. (1994) *On the Subject of Java*, Ithaca: Cornell University Press

Persoon, G. (1992) 'From sago to rice: changes in cultivation in Siberut, Indonesia', in E. Croll and D. Parkin (eds) *Bush Base, Forest Farm. Culture, Environment and Development*, London: Routledge.

Pratt, G. (1993) 'Reflections on poststructuralism and feminist empirics, theory and practice', *Antipode* 25, 1: 51–63.

Raliby, O. (1985) 'The position of women in Islam', *Mizan* 2, 2: 29–37.

Rosario, S. (1994) *Purity and Communal Boundaries*, Sydney: Allen and Unwin.

Sangren, P.S. (1995) ' "Power" against ideology: a critique of Foucauldian usage', *Cultural Anthropology* 10, 1: 3–40.

Saptari, R. (1995) 'Rural women to the factories: continuity and change in East Java's kretek cigarette industry', unpublished PhD thesis, University of Amsterdam, the Netherlands.

Schrijvers, J. (1988) 'Blueprint for undernourishment: the Mahaweli River Development Scheme in Sri Lanka', in B. Agarwal (ed.) *Structures of Patriarchy: The State, the Community and the Household*, London: Zed Press.

Scott, J. (1990) *Domination and the Arts of Resistance: Hidden Transcripts*, New Haven: Yale University Press.

Sen, K. and Stivens, M. (eds) (1998) *Gender and Power in Affluent Asia*, London: Routledge.

Siegal, J. (1969) *The Rope of God*, Berkeley: University of California Press.

Smyth, I. (1993) 'A critical look at the Indonesian government's policies for women', in J.P. Dirkse (ed.) *Indonesia's Experiences Under the New Order*, Leiden: KITLV Press.

Tirtosudarmo, R. (1995) 'The political demography of national integration and its policy implications for a sustainable development in Indonesia', *The Indonesian Quarterly* 23: 369–83.

Townsend, J., Arrevillaga, V., Bain, J., Cancino, S., Frenk, S., Pacheco, S. and Perez, E. (1995) *Women's Voices from the Rainforest*, London: Routledge.

Tsing, A. (1993) *In the Realm of the Diamond Queen. Marginality in an Out-of-the-Way Place*, Princeton: Princeton University Press.

Widodo, A. (1995) 'The states of the state: arts of the people and rites of homogenization', *Review of Indonesian and Malaysian Affairs* 29: 1–35.

Wieringa, S. (1988) 'Aborted feminism in Indonesia: a history of Indonesian socialist feminism', in S. Wieringa (ed.) *Women's Struggles and Strategies*, Aldershot: Gower.

Yuval-Davis, N. (1994) 'Identity politics and women's ethnicity', in V. Moghadam (ed.) *Identity Politics and Women: Cultural Reassertions and Feminisms in International Perspective*, Boulder: Westview Press.

# 6 Gendered surveillance and sexual violence in Filipina pre-migration experiences to Japan

*Nobue Suzuki*

[R]oots always precede routes.

(Clifford 1997: 3)

The notion of the diasporic ... concerns not only tracking the fluid movements of populations and of communities of affect, but also the violence that crystallises and propels such movements. Equally important, it also inquires into the machineries of disavowal ... that accompany such movements as well as attempts to combat such disavowals.

(Rafael and Abraham 1997: 150–1)

This chapter inquires into the pre-departure experiences of Filipina transmigrants married to Japanese men and currently living in urban Japan. My aim is to retell the stories Filipinas narrate about themselves and their own experiences of gendered surveillance and sexual violence, which led them to leave affective ties in the Philippines and move to Japan. The focus on pre-migration experiences provides a different vantage point to view the transmigration of Filipinas, adding to the growing literature on overseas Filipinas which has provided insights into these women's lives abroad and their ambivalent return visits to their 'homeland'.

A number of studies have highlighted the cultural rhetoric of leaving 'for the sake of the family' cited by many Filipinas as their key motivation for migration (for example, Lauby and Stark 1988; Trager 1988; Beltran *et al.* 1996; Sellek 1996). Trinidad Osteria (1994), for example, identified the ailing Philippine economy and the lack of employment opportunities as the primary reasons Filipinas give for taking jobs abroad. Using the 'family strategy' approach in analysing women's domestic migration, Osteria (1994) also argued that when women migrate, they do so to try to fulfil their financial obligations to their families. Studies taking such a perspective are clearly congruent with the position of the Philippine state which, since the beginning of the Aquino administration in the late 1980s, has officially proclaimed citizens involved in international migration as 'new hero(ine)s' (Tadiar 1997). More recently, President Gloria Macapagal-Arroyo called Filipinos in Japan 'overseas Filipino investors'.[1] These 'heroines' and 'investors' have been celebrated specifically because the benefits of their movements have flowed back into the national economy as various forms of remittance.

Transposing state rhetoric and rationale framing the international migration of its citizens on to the dynamics of family economics, however, unwittingly privileges state institutions while reducing migrants' individual agency. While the economic dimensions of migration are undoubtedly important to Filipina transmigrants, it should be noted that some women actually feel compelled to leave their natal families and marriages in order to work abroad and/or to (re)marry (unfamiliar) foreigners so as to get away from restrictive or distressing affective ties at home in the Philippines. However, compared to the focus on the economic logic of migration, this aspect of women's migration has thus far been given much less attention. My aim here is to show how particular cultural configurations and social relations in which pre-departure experiences take place – which Clifford (1997) refers to as 'roots' – bear on the 'routes' (i.e. courses through which migration experiences unfold) taken by these women. Central to my discussion are the ways in which women who have variously been troubled by the gender and sexual norms of their society actively seek alternatives that will enable themselves to find better life chances than before, even if it means they have to geographically and socially displace themselves from their natal families.

Observers of Filipinas in Japan have suggested that the women are multiple victims of gender, class, ethnic and national oppressions at home and in the host society, unable to challenge the situations in which they find themselves (Yamazaki 1987; David 1991; de Dios 1992; Yamashita 1995; Matsui 1997; Douglass 2000). In other instances, they are portrayed as opportunistic desperados reacting to such oppression by grabbing every chance of obtaining visas to facilitate flight or by chasing after material gain (see Suzuki 2000a: 149–55 for details). In this latter case, they 'marry or moonlight' only 'for the yen' (Sellek 1996: 159). These observations, however, also tend to present women's lives in terms of a single dimension. While poverty and the socio-economic mobility of the family are important reasons for migration in the Philippines, it should be noted that economic motivations are not clearly separated in the migrants' minds from other equally significant, non-economic reasons to move (Pertierra 1992: ix).

Beyond economic analyses, the growing literature on Filipinas in diaspora has provided rich insights into the women's lives overseas and their ambivalent transnational relations with their 'homeland' and the people left behind. While the continuing need for remittances may provide an important reason for remaining abroad, some Filipinas also find that their experiences of gender and sexual politics multiply the ambivalence of their relationship to 'home' and sometimes keep them away from 'home'. For example, some domestic workers in Hong Kong have found that during their absences their husbands began having affairs, while other women have developed a sense of independence which makes it difficult to bear their husbands' queries about their movements outside the home (Cruz and Paganoni 1989; Constable 1999). By crossing national boundaries, women abroad are often presumed to have also transgressed sexual boundaries (Constable 1997; Suzuki 2000b). In particular, returning women migrants from Japan today have routinely been identified as '*Japayuki* (Japan-bound)', a term which alludes to some kind of 'prostitute', working at bars or marrying Japanese men 'for convenience' (see Suzuki 2000a: 144–5, 149–52; Suzuki 2000b: 431–4, 439). Through the use of the 'euphemism' 'entertainers', the nature of their lives and work is often conveniently assumed rather than examined in its full complexity. Such discursive containment has instilled ambivalence about their home nation in the minds of many Filipina wives of Japanese men. It has also led them to engage in transnational projects of re-identification with the good Catholic Filipino wife (Suzuki, 2000b: 436–40).

Reflecting these unstable national identities and relations, Filomeno Aguilar (2000) questions the affective link between the state and the nation in the contexts of global emigration. He maintains that due to contradictory state policies that construct emigrants as vicious unpatriotic betrayers as well as virtuous 'hero(ine)s' of national development, migrants have come to realise that while their economic contributions aid nation-building in the Philippines, they also alienate them from the nation-state. Thus they value their 'Filipinoness' but not the Philippine nation-state. 'Home' does not eternally remain their affective anchor but becomes a locus of contestation of interests among people variously positioned in the global arena.

These emerging studies of ambivalent social relations have enriched understandings of the impacts and outcomes of Filipinas' transnational movements. They demonstrate that old forms of social relations come under question after Filipino migrants make contact with new sets of social relations in the course of their travels and displacement. Yet such studies have thus far provided little in terms of elaboration as to the reasons why, *prior to* their departure, some women decide to leave their families in the Philippines. Under ambivalent political–economic and gendered conditions, some women actually 'leave their homelands as a *wilful strategy to distance themselves from pained memories*' (Aguilar 2000: n.p., emphasis added). As I will show, contrary to the valorised notion of 'strong family ties' in the Philippines, the gendered surveillance and sexual violence to which some women are subjected within a broadly conceived 'home' propel them to reconfigure their lives through flights abroad.[2] What we need to examine, then, are embattled *gendered* histories and encounters that are constitutive of the intertwined roots/routes of the women's displacement (Clifford 1992).

## Filipina migrants to Japan

Living in a migration culture with Filipinos scattered over 130 countries (Gonzalez 1998: 39), many Filipinos imagine that life chances can be 'better' outside their home country. The early waves of Filipina migration to Japan in the 1970s began when contacts were established with Japanese businessmen and marital and sexual liaisons led some women to accompany their partners to Japan. Subsequently, others went as part of the Philippine (local) government-mediated labour and marriage migration schemes (Suzuki 2000a).

Since 1972, when the Marcos administration imposed martial law, Japan has become one of the major destinations for Filipinas. Due to Japan's strict immigration policy, the majority of Filipino entrants – aside from spouses, students and professionals – enter the country on 'entertainer visas' which are granted to musical, theatrical, athletic and other performing artists. These entertainers operate as bar hostesses, singers, dancers and sex workers, taking advantage of the many opportunities and higher wages available in the entertainment and sex industries as compared to those in other menial jobs commonly open to new migrants to Japan. As their work necessitates dealing predominantly with men, some of these workers subsequently marry Japanese nationals. Other Filipinas who marry Japanese through conventional and matchmaking meetings in the Philippines enter Japan as spouse entrants. Indeed, Filipina–Japanese marriages occupied first place among all filed intermarriages in Japan between 1992 and 1996.[3]

By 1999, following Koreans (636,548), Chinese (294,201) and Brazilians (224,299), Filipinos (115,685) comprised the fourth largest registered foreign population in Japan. Of these Filipinos, 84.8 per cent were women (98,103) (Ministry of Justice 1991–2000). Statistically, five groups – not distinguished by sex – comprised 89.3 per cent of the registered Filipino population in Japan in 1999: 'spouse-or-descendants' (this is treated as one category) of Japanese nationals (46,152 or 39.9 per cent); 'entertainers' (27,020 or 23.4 per cent); 'permanent residents' (14,884 or 12.9 per cent); 'long-term residents' (10,181 or 8.8 per cent); and 'tourists' (5,088 or 4.4 per cent) (Ministry of Justice 1991–2000).[4]

Of these groups, permanent and long-term residents are likely to be current or former spouses (and children) of Japanese citizens, while new entrants with 'spouse-or-descendant visa' holders are likely to be spouses.[5] One often hears it said among Filipino residents in Tokyo that *many* Filipinas in Japan either bring their children with them from previous marriages (or sexual liaisons) with Filipino men or leave these children in the Philippines. Osteria's (1994: 27) study of 155 entertainers supports this: of the 155 women, twenty-eight (18 per cent) were married, widowed or separated. Approximately 30 per cent of the Filipinas in my own research were previously married or had had live-in relations and 24 per cent had brought their children from the Philippines, despite the importance of virginity and lasting marriage under the Catholic doctrine during the time my subjects – now in their thirties and forties – grew up.

Bautista (1997) found that numerous young unmarried working-class mothers in Manila are subject to public surveillance of their female chastity and that they often suffer severe ostracism, poverty and reduced chances of marriage if they fall outside sexual norms. One rape survivor told me that she could not think of marriage while still in the Philippines, only finding relief after meeting her Japanese husband (in the Philippines) who showed more 'tolerant' attitudes towards non-virgins. Thus, some Filipinas' marriages to Japanese nationals appear to be a 'method' to undo injustices experienced in their home country and constitute one of the major motivations for their migration decisions.

## Women and agency

Filipino women's tales do indeed reveal that many leave home because of difficult and sometimes tormented relationships with Filipino men – fathers, brothers, husbands and sons – and often with other women complicit with the androcentric gender-sexual social organisation in the Philippines. This androcentric social configuration exerts two strands of control: one through the overarching surveillance of unwed women's virginity, and the other through the overstepping of sexual boundaries by men in the form of extramarital affairs and rape. The surveillance discussed here is articulated in four ways: the domination of fathers which goes against women's youthful desires; the scrutiny of women's sexual purity through gossip in the community; the infidelity of husbands; and domestic rape.

While all the stories below show their own idiosyncratic development, each woman's experience variedly overlaps with those of the other women described here as well as other women I know in the Tokyo and Nagoya areas. Furthermore, each of the four types of surveillance does not necessarily occur as an isolated, single incident; women's struggles may be compounded by multiple forms of surveillance and violence. These situations suggest the continuity, rather than the separable quality, of the gender and sexual control of women's behaviours, and demonstrate the complex ways in which the women are subjected to gender and sexual politics prior to their embarkations abroad.

Despite the hardships, women are still able to negotiate their gender and social relations (Clifford 1994: 313–14). Initially helpless and devastated, with time many of the women have been able to turn their misfortunes into opportunities and to recreate their lives (Ong 1995; Suzuki 2001). This may entail a risk or 'gamble' because the women leave their families and familiar surroundings to venture into new geographic territory and new relationships where social networks are virtually non-existent (Aguilar 1999). Still, the 'gamble' is taken for the opportunities thereby afforded to the women to improve their life chances. Fatalism is mediated by the belief that fate can be circumvented if creative actions are pursued (see Aguilar 1994: 151). The lives of women described in this study demonstrate that there are neither absolute, timeless 'heroines' nor 'victims'. Rather, given their circumstances, women make tough compromises between living with unkind social – including gender and sexual – realities and exploring alternative social and economic networks that appear to enable them to chart different lives outside their usual habitats (see Appadurai 1996: 48–65).

The data on which this article is based were collected through interviews and participant observation of over a hundred Filipina migrants in the Tokyo and Nagoya areas at various times between 1993 and 2001. The research is part of a larger project that inquires into the politics of affection among differentially positioned Filipina wives, their Japanese husbands and the women's families in the Philippines.[6] Below, I draw on the experiences of eight women as fuller instances of the movements of Filipinas who had undergone gendered and sexual hardships prior to departure. Of these eight women, I have been in contact with five fairly regularly since 1993. I was introduced to the sixth woman in 1996 by her

Filipina church-mate (and while I did not get to know her very well, our conversation in Tagalog at her home seemed to ease her anxiety sufficiently for her to tell me her story). All interviews were conducted at the subjects' homes, except for one woman with whom I talked alone on the rooftop of the building where her bar was located. The language used was chiefly Japanese with four women and Tagalog with two. Two of these women related to me rape assaults experienced by their sister and niece. In the next section, I examine the sexual and gendered reasons behind these Filipinas' decisions to leave their families and communities.

## 'Strict' fathers

### Imee[7]

Imee is one of many Filipinas who went abroad to escape authoritarian parents (see Samonte 1992). She grew up in a fairly well-to-do family in Manila and has four older and three younger siblings. According to Imee:

> There're many good things at home, but what wasn't nice about my father was that he was *veeeery* strict towards us. My eldest sister eloped when she was 16. Soon after that, my brother had a forced marriage. He brought his girlfriend home and she stayed until midnight. He was 17. My two other sisters also married early and left home. I think this was because of my father's strictness. Then, my father's sternness was directed at me. When I started to work, how late could I stay outside? For sure, I came back by seven or eight. He ordered a curfew. Just once, I came back late at about eleven. I was with my brother but my father was infuriated!

When Imee was still in college, her father suddenly stopped working, and as the eldest child at home Imee became the breadwinner of the family. She had an office job during the day and she sang at a hotel in the evening. She said,

> I didn't think of the singing job [as one I did] for money. It was for fun and freedom. I fought against my father's strictness. I didn't listen to him because this is *my* life. I wanted him to let me do what I want to do.

Apart from her workplaces, Imee's movements were restricted to her home, church and shopping malls, to which she went with a chaperon.

At the age of 20, considering herself 'too young to worry about marriage', she began exploring alternatives for her own life. Although she was engaged to a Filipino pastor-to-be from her church, she could not marry him immediately. She also had a male American pen-pal and mentioned that a close friend had married an American man whom the friend had met through correspondence. Subsequently, Imee met a Japanese at the hotel where she worked as a singer and was attracted to his intelligence. She felt that she 'couldn't wait for [her] fiancé to complete his training in another year' because she felt stifled by her father's curfew and chaperoning. She commented:

> Life is a game [*shōbu*] usually. Of course, I play it with much thinking. At first, I didn't have any plan of marrying him [the Japanese boyfriend], but I thought of the practicality of it. Love developed afterwards. My father said, 'Don't you think of marriage so casually!' [I replied,] 'I'm not thinking of it casually. I've thought about it a lot [meaning "seriously"]. Please meet him. If you say, "No", I'll obey your words'. As soon as he met my husband, my father was impressed by his intelligence and seriousness.

By planning her 'play' in the 'game', Imee successfully negotiated with her father about what she wanted for her own life. Although Imee initially denied any possibility of marriage, she expanded her life chances by having an admirer from abroad, and thereby achieving some sense of autonomy from her father's tethering.

### Millie

For some Filipinas in Japan, paternal strictness may be compounded by other problems. Millie is the child of a constabulary officer and his mistress. She has an older half-sister and -brother on her maternal side, but she lived in her father's home. Although she grew up enjoying a comfortable life, she also had a chastity-demanding, 'militaristic' father (despite the fact that he himself was a womaniser) who rarely allowed his children to go out, even with friends. She explained:

> Every day he drove us to school in the morning and picked us up in the afternoon. He also knew my schedule and when my last class ended at four, he had come to school by three-thirty. We couldn't go anywhere! He even drove us to church, which is only a three-minute walk from home! I felt like I had no human association except with my father. He used to tell us, because there are many politicians in his family, 'Don't disgrace my name!' If one of his daughters did not marry, if they gave birth to a fatherless baby, like that [*sic*]. But when I got older, I gradually felt bored at home.

It was not just 'boredom' that Millie felt. Her status as the child of a mistress complicated her relationships at home. Millie explained:

> I lived with my half-siblings. Sometimes, my father quarrelled with his wife. When my father was absent, my half-sibs told me, 'Because of you, our parents fight against each other'. I wondered why *I* was the reason for their problems. So, I left home once [when she was in her mid-teens]. I went to my brother's home and he sent me to high school. Because I wanted to go to

college, I worked as a campaign girl. Then, my father came and 'arrested' me. I didn't like it! I had already experienced life outside and so I didn't want to go back to the gloomy life at my father's home. 'Don't do this! Don't do that!' I hated it! [While I was working,] I had a boyfriend. I was 16 or 17. Then, I got pregnant, but he asked me to abort the child. I thought he wasn't a man. No way! [Later,] my father supported me with some money. I realised that he was concerned about me. I thought that perhaps my father also suffered, but after all, he was at fault. He made children one after another.

A few years after her son's birth, Millie decided to go to Japan as a dancer-entertainer, capitalising on the dance training she had as a teenager. For her, this presented not only a 'fun' opportunity but also one that allowed her to give a good life to her child and to her maternal family who had supported her after she left her father's house.

She is currently married to a well-to-do Japanese businessman who built a large house with some 300 square metres of floor space in an exclusive subdivision in central Manila. Millie moved her child, her mother, her brother and his children into this comfortable home. All the children in the household are sent to private schools. With her substantial financial contributions of ¥150,000 (US$1,200) in monthly remittances, she has gained much respect in her natal family. The once bullied child of a mistress now gets invitations from relatives whom she has helped, to attend important events such as weddings and baptisms. While in Japan, she puts up with the life of an isolated urban housewife, but in the Philippines she is the 'President' of her imposing home where people gather to please her whenever she visits. Given these material situations as well as her father's passing several years ago, Millie has now displaced, within her household, the sexual surveillance and status hierarchy from which she suffered, although she, like other women, does not subvert the totality of such control and domination in Philippine society.

## Communal surveillance

### Lily

In the Philippines, the massive public circulation of rumour and gossip exerts political power (see Rafael 2000). Rumours and gossip substitute for the officially subscribed version of the way things are (Rafael 2000: 116) as well as reinforce how things ought to be. They are thus integral to gender and sexual politics and work to circumscribe women's sexual boundaries to a single partner within marriage. As in most societies, men's marital choices, especially for men of higher social strata, are also subjected to such informal scrutiny. Controlling young women's sexuality matters because it is a way for men to demonstrate their power and authority in gender-sexual politics, serving to strengthen the androcentric social system (Ortner 1996: 43–58).

Lily suffered from such control in the form of gossip, which undermined her future marriage prospects as a young woman. As I go on to show, the communal doubt cast by gossip on her female 'virtue' was the primary cause that led to her moving out of her natal community. After attending college for three years, she had to find work in a factory to support her family because her father fell ill. She was successful at factory work, but she was 'not successful in [her] love life'. She elaborated on her experience:

> At 17, I had a boyfriend. But our relationship did not work out. Then, at the company, I got another boyfriend. My new boyfriend learnt about my ex-boyfriend through gossip! We were all in the same company. You know, Filipinos, if possible, they like virgins there, right? Once you have a boyfriend, it's like, 'Oh, you have already been experienced. You can't be helped'. I was depressed. As soon as he learnt about [my old boyfriend], he lost respect for me. [He said,] 'You really knew him, huh?' Even when I went to a movie with a [girlfriend], he didn't trust me. He thought I was meeting my old boyfriend. Then, he married another girl.

The gossip about her first failed love life spread throughout her workplace, where Lily had met her first and second boyfriends. Gossip of this kind that cannot be verified (Rafael 2000: 117) permeated into even the tiniest space surrounding Lily and circumscribed her movements as well as choices in life. She felt that if she had stayed on at the same workplace, she would have been afflicted by the proliferation of gossip had she tried to start another relationship. She was eventually compelled to leave her job and look for other means to sustain her life, a difficult task in a context where unemployment and underemployment for women are especially high (Eviota 1992). Lily explained how she attempted to untangle herself from such communal surveillance and disparagement:

> Now, I thought, how do I carry on working? [But] I need money to live! Really, I lost any chance to have a happy life in the Philippines. That's the difficulty in the Philippines. It's women's fate there … Then, I had a friend whose cousin was married to a Japanese and they introduced me to a Japanese man [whom Lily eventually married] … When I got married, it was like suicide. Whatever awaited me, I just wanted to leave my country! Now, I feel better because I have adjusted to Japanese life. [Reflecting on the irony of her circumstances, Lily added the following:] Since I was a child, I had told myself, 'When I marry, my life is for my family!' Now it is different!

The combination of fear of communal gossip and financial hardship led Lily to choose what she regarded as a 'suicidal' route to Japan even as she affirmed her determination 'not to return to my community should [my] marriage fail!'

In her living room, Lily displays a framed photograph of her family at Manila International Airport receiving her coming back from Japan carrying large pieces of luggage full of souvenirs. Lily is shown wearing a gorgeous smile as if to celebrate her victorious homecoming. Her initial fears that by marrying a foreigner she had chosen a 'suicidal' route which would lead to an uncertain future appear now to be unfounded. Ten years after marriage, Lily and her husband have purchased two lots in Manila and Laguna where they plan to retire after their children reach adulthood. These visible acts of material display in the Philippines seem to be a way through which Lily circulates counter-discourses against the disparaging gossip that had driven her from the Philippines. As Rafael (2000) shows, rumours and gossip create an imaginary space for negotiation with and resistance to overbearing forces. Material display hence provides Lily with a means to demonstrate that she has become a cultur-ally sanctioned wife and mother as opposed to the woman that she was once rumoured to be. Against the communal surveillance of her sexuality, they articu-late Lily's desire for personal dignity and social respect for her renewed status (see Aguilar 1999).

## Womaniser husbands

### Pinky

Amid the lurid accounts of the life of domestic worker Flor Contemplacion, who was executed by the Singapore government for the murders of her employer's son and a fellow domestic worker in 1995, it transpired that Flor had gone to Singapore to heal the misery caused by her husband's extramarital affair (Torrevillas 1996: 46). Like Flor, Pinky also left the Philippines in order to leave a broken marriage and to distance herself from her womaniser husband.

As with Imee and Millie, Pinky also had a strict father who imposed a daily curfew at 5 p.m. As a teenage girl, Pinky wanted to enjoy herself, but her father's discipline prevented her from playing outside the home. When she was 15, she met a 19-year-old Filipino man with whom she eloped. One year later, she gave birth to her first child. Her own preference was to be a housewife, and she busied herself caring for her husband's and child's needs. However, her marriage did not last very long, and she was pushed into an emotional abyss:

> When he was about to leave home [for work], I polished his shoes. I prepared bright, white underwear. If it's the man I love, I'm like that. Then, he went drinking for his business ... That made the situation bad. He started to have affairs. How many times? He came back in the morning, having kiss marks. He smelt different. I knew it! He had spent time at a 'love hotel'. Once we were separated and his parents came to ask me to stay with him. He too cried and begged for forgiveness. He was still young, so I accepted it. But ... I later learnt that he and his girlfriend had had a child. I couldn't forgive him any more!

In the meantime, Pinky gave birth to her second child. On learning about Pinky's circumstances, her older sister, who was married to a Japanese, invited her to Japan. During her second stay, she was introduced to a Japanese man whom she 'dated' for a few weeks with the help of interpreters. Although Pinky herself was unsure about the man, her sister insisted that Pinky start a relationship with him as a means of renewing her life in Japan. As Pinky had been under-age when she first started living with the Filipino man who fathered her children, their marriage was not registered despite promises on his part to do so. Ironically, this situation enabled her to marry the Japanese man.

Although Pinky appreciates many things that her current husband has done for her and her children, whom he adopted, she confessed that she in fact continues to love her Filipino 'husband'. Her older daughter looks just like her father, reminding Pinky of him. She related the ambivalence she felt whenever they meet:

> I meet him even now. Of course, [since] he is the father of our children. Frankly, I love him only and I'm sorry [for my present husband]. But [my ex-husband] lives with another woman and their children. When my older daughter had her debutante, I held a party for her in the Philippines. I didn't tell my [Japanese] husband about this but I invited her father. Well, they are father and daughter. We sang and took pictures. After the party, everyone slept in a large bed. I stayed away from him … He embraced me tight. I said, 'No, we can't go back'. He only then told me, 'I love you'. [I replied,] 'You silly! I no longer have feelings for you'. [Interviewer interjects, 'Really?'] Yes, I do [continue to have feelings for her ex-husband], but I must be firm. He can't draw the line. But I was happy that he embraced me.

Obviously it was not only her ex-husband who had to suppress feelings of continued love and affection. In fact, since Pinky remarried in the early 1990s, the children have been her reason to continue meeting her former husband.

Pinky augments her income by working as a hostess at an exclusive night-club in Japan. She has been doing so for the last six years. She noted that such work enables her not only to acquire proper linguistic skills for her life in Japan but also to learn about how Japanese men run their businesses (since many of her clients are businessmen). Although dates with clients are often part of a hostess' job and there are men who think of her as their 'girlfriend', Pinky denies sexual engagements with her clients. She rationalised:

> Yes, I go out with them to have a meal and a talk. But it's not sexual. I have only one body and I can't satisfy everything. What's more, if our relations are sexual, when they get bored with me, that's the end. I can't have lasting clientele for my business once I engage in sex with them.

Pinky's first marriage has left her with painful lessons about managing relationships with men. At the bar, she is attentive, listening to her clients (see also Allison 1994) and gains knowledge about how to do business with the Japanese. Her original wish to be a housewife who prepares bright shoes and clean underwear for her husband has not materialised. Instead, her life trajectory has taken an unexpected turn. Pinky has learnt a great deal about how to deal with men. She is now a businesswoman who works to enjoy a new life in Japan but also to appease the ambivalent sentiments she still has about the life she left in the Philippines.

### Rose

Rose was a working student when she married her 'childhood sweetheart' at the age of 23. With her university degree, she landed a clerical job at a research institution and her husband, a university dropout, began working as a security guard. Rose also engaged in various kinds of labour- and consumer-related activism in the Philippines. Two years after marrying, she suffered a miscarriage. Because of the hospital bills, the couple went into debt, leading to the dissolution of their marriage. Rose recounted:

> We spent lots of money. Then, my husband said, '*Mag-abroad na lang* (I'll just go abroad)'. That was his ego because of our incomes; mine was higher than his. He has his pride. He was wholly a *man*. Time had just come. He became a security guard at Jeddah International Airport. He sent me $100 out of his $250 monthly salary. Then, my husband stopped writing me. Suddenly, we lost communication for about three years! So I told my siblings that I too will just go abroad rather than waiting for nothing. We have relatives in the States. I had my brother and cousins in Japan. One of the cousins is married to a Japanese [in Japan]. So, they said, 'Come to Japan!' But before I went to Japan, my husband came back. We met and he was still passionate about me. At this time, he had another woman. She had also come back from Saudi. He had telephone numbers, grapes, chocolate, soaps and things. Those were for her souvenirs. Oh Lord! 'Yes, I have [a woman],' he said. I couldn't do anything. My awareness [about various social issues led me to think] that I had become a victim of the system of our society. All these are effects of overseas migration!

In 1992, Rose joined her brother and two cousins in Japan. Her relationship with her husband waned.

Rose picked up several jobs and worked day and night, overstaying her student visa, until she got married to a Japanese.[8] After arriving in Japan, Rose learnt that one of her cousins there was extremely stressed as a result of the total dependence of his family in the Philippines on his earnings. One day, he jumped from a high building and damaged his spinal cord, leaving him unable to work. This incident, together with her own bad experience with a womanising ex-husband, revived Rose's desire to raise awareness about the plight of her countryfolk in Japan. Being childless helped Rose win her Japanese husband's understanding about working as a hostess in order to raise funds for her projects. Rose is so committed that she now enjoys the support of local Japanese activists and lawyers working for the empowerment of her *kababaihan*, or fellow country-women. Rose is also involved in feminist-activist organisations in the Philippines so as to be able to link the voices of marginalised migrants in Japan, especially women, with gender and class-based oppression in the Philippines.

## Rape

### *Farah*

Farah lost her father when she was 12. Consequently, she joined the labour force by helping her mother make crafts for sale. While working at home one evening they ran out of materials, and Farah had to go to the factory alone because her mother was suffering from a headache. Her life changed after this:

> So I went there by myself and a man also working at the factory was waiting for his chance. He had a crush on me, but I didn't know about it because I was only 13. When I was about to leave [the factory, I was raped] … I conceived my first son. [Two years later,] I conceived my second son [by the same man, with whom she had developed a relationship]. Then, the man disappeared. My siblings were ashamed of me and [one of] my brothers' wives especially bullied me. When I went home, she spat and said, 'The pig's come back!' I had two fatherless children, so I was a 'pig', she said. That hurt me! So, every Saturday night, I spent time at a disco. I came home in the morning. I slipped into my bed quietly while everyone was still sleeping.

At one of her excursions to a disco in the early 1980s, Farah, then in her mid-twenties, was recruited to go to Japan as a dancer. Soon after she began making trips to Japan, she conceived another child. Her Japanese partner promised to marry Farah; however, after learning about her two teenage sons, this man also disappeared. At this time, she decided to leave home in the Philippines and move to Japan. She explained: 'I didn't want to be killed. Well, I didn't want to be bullied again'. In the mid-1980s, Farah married a different Japanese man who adopted her last child. Because the records showed that she was an illegal entrant, she was not granted a spouse visa for one year despite being formally married to a Japanese. In the end, Farah failed to bring her two older children into Japan before they reached the age of 20, the age at which the Japanese government no longer regards children as 'dependants'.

Since she began working in Japan, Farah has continued to support her sons. Without parental supervision, her sons developed unfavourable habits. Her first son, Gary, began hanging around with his high school buddies and learnt to use drugs. He dropped out of school and later went to Japan, where he worked illegally in order to support his newly formed family in Manila. Farah's second son, Orly, finished university; according to his mother, he is 'proud of his intelligence' but he 'does not seem to be seriously looking for a job'. Farah voiced her regrets: 'I spoilt them. I let them enjoy a good life. I must give them pains that cut to the marrow!'

At the time of the interview in 1996, not only had Gary overstayed his visa, he was also drinking, using drugs and having an affair with a married Filipina in Japan. His lover called Farah a 'beast' when Farah asked her to stop seeing Gary. While taking me back to the train station, Farah pointed to this woman's club and said, 'If I see my son now, I'll take him to the police and tell them, "Please arrest and deport this son of mine!"'

Because of the prolonged recession in Japan in the 1990s, Farah and her family in Japan have been facing financial difficulties. Still, she manages to make a trip back to the Philippines at least once a year. She goes back partially because 'I want to complain to everyone!' The bullying sister-in-law has now become financially indebted to Farah and therefore no longer throws vicious words at her. Farah also enjoys the respect of her neighbours whom she gives small gifts – candies and ribbons. Their words of appreciation seem to be one way in which Farah sustains her ambivalent link with people in the Philippines.

### Linda and Carmy[9]

Rape violates not only young virgins but also married women. Linda is Imee's sister. In order to leave her strict father, Linda 'married' a military officer. By the time she learnt that he was already married with two children, she had had a child with him. The situation turned violent when he held a gun in his hand and raped Linda. She conceived her second child as a result. After this traumatic incident, she left the man and the Philippines and began working at an inn in Japan. There she met a Japanese and married him. She had two children by this husband, and he adopted her older child. Linda did not tell her Japanese husband about her second child who is being taken care of by her mother. According to Imee, Linda avoided taking the child to Japan since she did not want to be reminded of the rape assault.

Another victim of rape is Rose's niece, Carmy, who was raped by her own widowed father. Carmy had wanted to commit suicide, but eventually Rose, whom she consulted, changed her mind. According to Rose, although Carmy's boyfriend had been very understanding, Carmy's fear was that, as a man, he might do the same to their daughter when she grew up. Carmy has expressed her desire to join Rose in Japan, bringing her daughter along; if not, she has asked that Rose adopt her child. Japan's immigration laws, however, are very strict about issuing even 'family visit visas' to young family members of Filipinos, given the (unofficial) national policy of keeping the population ethnically 'homogeneous' as well as suspicion that 'foreigners' will engage in illicit work such as prostitution and bar-hostessing. As an activist, Rose knows too well that it is virtually impossible to bring her niece over. Though the Japanese government has just begun to develop a scheme which will allow professionals to work in Japan (*Migration News* 2001), it is unlikely that people like Carmy, even with a pharmacist's licence, will be granted a visa enabling them to work and bring their families to Japan for good. Thus Carmy continues to seek a ticket not only for a flight out of the Philippines but also for empowerment as a woman and as a professional. At the moment, Rose can only tell Carmy to endure the present hardship or to try to process papers to go, as a pharmacist, to the United States, where they have relatives.

## Gendered boundaries, sexual oppression and migration

The stories of the women introduced here illustrate how different forms of sexual politics and violence have disciplined their lives to various degrees and over different spans of time. As much as parental curfews and supervision curtailed the movement of Imee, Millie and Pinky, gossip also put pressure on Lily. The total effect was to push these women into unmooring themselves from the restrictive norms they felt in their natal community. All the subjects in this essay considered the norms restrictive and tried to expand their life chances by navigating a potentially 'suicidal' voyage across the unfamiliar sea of migration. What the women's experiences have also shown is that the high incidence of women with children from previous liaisons travelling to Japan has its partial roots in the sexual politics back home in the Philippines.

In contrast to Filipinas, Filipino men appear to have a far greater degree of freedom and can overstep the sexual exclusivity of the modern Catholic marriage: womanising fathers, husbands and sons have caused their daughters, lovers, wives and mothers much agony. Millie's authoritarian and promiscuous father's attempt to control her virginity created adverse effects in Millie's heart in her early life. Extreme cases of male sexual freedom turned into bigamy and led to the violent rape assaults on Farah, Linda and Carmy. While Farah has gained some power to 'complain to everyone', the stability of Linda's new life is maintained only by the continual hiding of part of her past, and Carmy thus far sees no exit from her trauma.

Gender politics has been a crucial mechanism at work. Millie's father was concerned about his name in relation to his politician relatives. For men, especially those with higher social status like Millie's father, mutually constitutive male honour and power demand the control of unwed daughters' virginity (see Ortner 1996). As the head of a patriarchal family, he is 'not simply responsible *to* his family but also *for* his family *vis-à-vis* the larger system' (Ortner 1996: 53, emphasis in original). Conversely, Rose's working-class husband had been frustrated because he did not have a university degree – an important symbolic capital in the Philippines – or a higher income than his wife, and took his chance to declare, 'I'm going abroad!' in order to assert his male ego.

What is ironic about these gender struggles is that Filipina migrants like Farah see other women in the Philippines as complicit in the conspiracy that constrains women. Farah's suffering was compounded by her sister-in-law's verbal assaults which caused her to feel as though she was being 'killed'. Forced to become sexually active at an early age, Farah was morally and psychologically isolated from family members for over ten years despite her physical co-residence with them. This rape survivor and single mother found herself, to borrow Constable's (1999:

203) phrase, 'at home, but not at home' in her gender-sexual exile, *even before* she was compelled to find a way to sustain her family outside the nation. Farah's consequent migration abroad was furthermore hardened by her son's lover.

Despite the pervasiveness of gendered surveillance and sexual violence at home, the women under discussion have variously resisted the oppressive systems because, in Imee's words, 'This is my life'. While overseas flight does not easily offer emancipation, these women have nonetheless empowered themselves by reading and translating domination differently in diverse arenas of their social experiences (see Ong 1995). They have strategically made use of the resources, especially human capital, available to them to remap their agency at home and in diaspora (Ong 1995).

The women's strategies varied from very subtle manoeuvring to blatant rejections of the norms regarding sexual purity. As they began to expose themselves to the world outside the home and out of the reach of their domineering fathers, Imee and Millie in particular enabled themselves to resist the paternal tether. Imee performed the role of an obedient daughter and listened to her father carefully: 'If you say, "No," I'll obey your words'. From this subordinate position that exalted her father's paternal power, she nonetheless successfully steered his mind to agree to her marriage. For Imee, Lily and Pinky, culturally sanctified marriage was a legitimate and 'practical' means to deflect the pervasive sexual control. Perhaps for them this was a different kind of 'family strategy' for liberation from surveillance at home.

Going against the wills of her 'militaristic' father and 'unmanly' boyfriend, Millie became a teenage single mother. Millie first escaped from her constabulary father's 'arrest'. She then reactivated her once-disrupted relationship with her maternal half-siblings and tapped into the strategic memories of kinship with her half-brother (see Aguilar 2000) in order to empower herself to transgress the normative behaviour expected of daughters. Subsequently she challenged the cultural taboo against single motherhood and rejected her lover's request to abort the child, to avoid subjecting herself to yet another set of controls.

By breaking the taboo, Millie also broke the hegemonic construction of the unwed daughter as virginal and dutiful, and laid out alternative moralities (see Ong 1995: 359). Her participation in the external world transformed her relationship with her tyrannical father and remade her agency, which in turn served as a form of critique against the overarching controls over women's sexuality (Ong 1995). Fortunately and somewhat paradoxically, in the course of her struggles towards self-empowerment it was her brother, a man, who supported her way out of the restrictive sexual regime.[10]

For decades, Filipino returnees from Japan have invested much in houses and land in the Philippines for their future security and as a form of redemption for their difficult lives abroad. While these enable Filipinos to show that they have successfully swum against the currents of uneven global capitalism, they also serve as reminders of their being not only economic pilgrims but persons with their own desires, dreams and dignity (Aguilar 1999). For women like Lily, Millie and Farah, tangible symbols such as land, imposing '*Japayuki* palaces', multiple vehicles, brand goods, numerous housekeepers, private-school education and even small gifts of ribbons and candies act not only to redeem personal satisfaction, but also to provide them with a voice to speak to their families and neighbours and to a nation which had disparaged them as 'pigs' and 'prostitutes'. Against their former identities, such status symbols are circulating contrapuntal gossip that the women have now become successful and are enviable role models (Aguilar 1999: 105). Their unexpected economic ascendancy, which has been realised through marriage, especially to foreign nationals,[11] unfortunately also generates a third set of cacophonic gossip full of jealousy.

Filipinas have also attempted to rewrite assumptions about their lives which often confine them to being agents in the Japanese 'sex industry'. Because of the overlooked plurality of Japan's entertainment and sex industries, Filipinas associated with Japanese men have been constructed as 'victims', 'opportunist prostitutes' or '*Japayuki*'. However, the work actually performed in night-clubs in Japan is far more complex (see Ballescas 1992; Allison 1994). The complexity becomes legible when women workers' agency is recognised without falling into the trap of moralist discourse. In fact, Pinky, for instance, has capitalised on the opportunities – beyond tangible tips, gifts and marriage proposals – that her clients and bar business offer to her. Bar hostesses in Japan do not always trade their bodies. Furthermore, as in the case of Pinky, countless Filipina entertainers have converted their assignments into strategies to learn the Japanese language, including quickly reading the various scripts on karaoke monitors. Learning the language and culture of the host society in this context does not imply unilateral assimilation. Rather, cultural knowledge thus attained becomes a useful tool to enable social and economic mobility. While Filipina entertainers have been tickling male egos and have sometimes been treated as playthings of male sexuality, others fit their work to their personal goals.

Being aware of the effects of sexual discrimination, global capitalism and their ramifications on migration in the Philippines, Rose, perhaps more than the others discussed here, has avidly made direct acts of resistance in an attempt to remake life in diaspora. Through her transnational activism, Rose reminds us of the presence of disadvantaged women and families living under oppressive gender-sexual regimes and uneven material conditions. Her activism exposes the mechanisms of oppression and disavowal faced by sexually and economically alienated women in both Japan and the Philippines.

## Conclusion

Living as members of a migration culture, many Filipinos have become 'gambler-migrants' who test the limits of fate through their pilgrimages abroad (Aguilar 1999). To differing degrees, Imee and Lily are risk-takers, as their comments that 'life is a game' demonstrate. They actively seek to maximise the opportunities of their encounters with foreign men. These women's 'gambling' then negates the notion that they leave their lives passively to chance. Rather, they are women who plan and take responsibility for their own life trajectories even as they are aware of the risks and uncertainties, as reflected in Imee's words, 'I play it with much thinking'. Others seek safer routes through transnational linkages and communications sustained by networks of overseas relatives, friends and even foreign pen-pals. Thus, Pinky, Rose and Linda were able to imagine a plural, deterritorialised world in which they can seek alternatives. It should also be noted that Japan is geographically close by, offering tens of thousands of jobs especially to women, despite associated potential dangers and dishonour. Millie and Farah thus joined the massive exodus of young Filipinas to Japan's entertainment and sex industries in order to overcome hardships faced in their home nation.

For the past few decades, Japan has thus offered a 'haven' for Filipino women who have suffered emotionally and sexually but are determined to overcome 'women's fate'. As in the metaphor of 'gambling', everyone begins the game of life with a vision of winning and not losing (Aguilar 1999). In participating in this game, some women have had to bear considerable 'handicaps' as well as the lingering effects of violence which continue to affect their lives even in diaspora. How successful they become is also contingent on how they negotiate – and the degree to which they become subjected to – new forms of surveillance and discipline in their new families and communities. Nonetheless, while in diaspora many Filipino women do find and build a new 'haven' in Japan where they can moor for good, or at least for the time being.

In this chapter, I have suggested that women's strategies for survival and propensity to move began while in the Philippines. To understand migration, we thus needs to expand our analytical time frame and situate women not only during and after the 'routes' of their movements but also in the 'roots' of their social relations at 'home' prior to embarkation. The women's tales narrated here also reveal that the valorised rhetoric of 'for the sake of the family' and 'family strategies' needs to be understood with caution and that women's struggles for emancipation may be rooted in affective associations in which androcentric gender and sexual modalities, disciplines and violence are simultaneously practised not only by men but also by women who are complicit in spinning the web of control around other women. This is one reason why women feel ambivalent about and sometimes disavow their 'homeland' and nation. Nonetheless, the women's pre-departure experiences have clearly demonstrated the dynamics of ongoing trajectories which they (re)create for themselves – from varying degrees of gendered grievances to different types of sexual defeats – through the forging and mobilising of their agency in order to strive for new life chances in an otherwise unkind world.

## Acknowledgements

An early version of this paper was presented at the conference *Migration and the 'Asian Family' in a Globalising World*, organised by the Asian Metacentre for Population and Sustainable Development Analysis, National University of Singapore, 16–18 April 2001. I thank the organisers and participants for the encouragement that I received. I also thank the Japan Foundation and Matsushita International Foundation for providing me with generous funds. Last, but not least, I would like to express my deepest appreciation for the courage of the six women described above who told me their own and their female kin's stories.

## Notes

1 Speech delivered by Gloria Macapagal-Arroyo, President of the Republic of the Philippines, at a reception held by the Philippine Government in Tokyo on 15 September 2001 and attended by about a thousand Filipino residents in Japan.
2 By saying this, I am not suggesting that such surveillance and violence are peculiar to the Philippines. Different forms of masculine social structures and practices also control women in host societies (Piper 1997; Suzuki 2000a).
3 Filipina–Japanese marriages thereafter occupy second place after Chinese women–Japanese men marriages (Ministry of Health and Welfare 1993–2001).
4 Descendants in the 'spouse-or-descendant' category are divided into two groups: second- or third-generation descendants of Japanese nationals who out-migrated or were born abroad (*Nikkei*); and Japanese nationals' children who do not possess Japanese citizenship. Long-term visas are granted to refugees and 'foreigners who "have consanguineous ties with Japanese society"' (Ministry of Justice Immigration Control Bureau 1998: 95–9). Permanent visas are not readily issued and marriage to Japanese nationals is one of the fastest ways to receive them.
5 The number of new entrants does not represent that of residents and this number is meant to show a tendency only.
6 This long-time relationship is partially the result of the fact that I have lived in the designated areas since 1995 with a one-year intermission between 1999 and 2000. Limited space does not allow me to fully explicate the relationships of the differentially positioned researcher and researched, which are actually much more complicated than sketched here.
7 All of my subjects' names are pseudonyms.
8 Divorce is not legally allowed between Catholic Filipinos. However, Rose said that because her husband was 'caught in the act', she was able to nullify her marriage.
9 The stories of Linda and Carmy were obtained from Imee and Rose respectively.
10 Although sibling ties are generally strong among Malayo-Polynesian people (Aguilar 2000), individual experiences may vary under different circumstances. Farah's tough life, for example, began after her father's death and when her eldest brother breached his promise to her teacher who had offered to send Farah to school, saying that he would take care of his sister. He kept his promise for only one school term.
11 Foreigners are not granted an equally 'high' status and Japanese generally come below 'Americans' (Ventura 1992).

## References

Aguilar, F. (1994) 'Of cocks and bets: gambling, class structuring and state formation in the Philippines', in J. Eder and R. Youngblook (eds) *Patterns of Politics and Power in the Philippines*, Tempe, Arizona: Program for Southeast Asian Studies, Arizona State University.
——(1999) 'Ritual passage and the reconstruction of selfhood in international labour migration', *Sojourn* 14, 1: 98–139.
——(2000) 'Ambivalent narratives of Filipino transnationalisms in the Asia-Pacific', paper presented at the *International Conference on Transnational Communities in the Asia-Pacific Region*, 7–8 August, Singapore.
Allison, A. (1994) *Nightwork: Sexuality, Pleasure and Corporate Masculinity in a Tokyo Hostess Club*, Chicago: University of Chicago Press.
Appadurai, A. (1996) *Modernity at Large: Cultural Dimensions of Globalization*, Minneapolis: University of Minnesota Press.
Ballescas, M.P. (1992) *Filipino Entertainers in Japan: An Introduction*, Quezon City: The Foundation for Nationalist Studies.

Bautista, P. (1997) *Young Unwed Mothers*, Quezon City: Giraffe Books.

Beltran, R.P., Samonte, E.L. and Walker, L. (1996) 'Filipino women migrant workers: effects on family life and challenges for intervention', in R.P. Beltran and G.F. Rodriguez (eds) *Filipino Women Migrant Workers: At the Crossroads and Beyond Beijing*, Quezon City: Giraffe Books.

Clifford, J. (1992) 'Travelling cultures', in L. Grossberg, C. Nelson and P. Treichler (eds) *Cultural Studies*, New York: Routledge.

——(1994) 'Diasporas', *Cultural Anthropology* 9, 3: 302–38.

——(1997) *Routes: Travel and Translation in the Late Twentieth Century*, Cambridge: Harvard University Press.

Constable, N. (1997) 'Sexuality and discipline among Filipina domestic workers in Hong Kong', *American Ethnologist* 24, 3: 539–58.

——(1999) 'At home but not at home: Filipina narratives of ambivalent returns', *Cultural Anthropology* 14, 2: 203–28.

Cruz, V.P. and Paganoni, A. (1989) *Filipinas in Migration: Big Bills and Small Change*, Quezon City: Scalabrini Migration Center.

David, R.S. (1991) 'Filipino workers in Japan: vulnerability and survival', *Kasarinlan* 6, 3: 9–23.

de Dios, A.J. (1992) 'Japayuki-san: Filipinas at risk', in R.P. Beltran and A.J. de Dios (eds) *Filipino Women Overseas Contract Workers: At What Cost?*, Manila: Goodwill Trading.

Douglass, M. (2000) 'The singularities of international migration of women to Japan: past, present and future', in M. Douglass and G.S. Roberts (eds) *Japan and Global Migration: Foreign Workers and the Advent of a Multicultural Society*, London: Routledge.

Eviota, E.U. (1992) *The Political Economy of Gender: Women and the Sexual Division of Labor in the Philippines*, London: Zed Press.

Gonzalez, J. (1998) *Philippine Labour Migration*, Singapore: Institute of Southeast Asian Studies.

Lauby, J. and Stark, O. (1988) 'Individual migration as a family strategy: young women in the Philippines', *Population Studies* 42: 473–86.

Matsui, Y. (1997) 'Matsui Yayori', in S. Buckley (ed.) *Broken Silence: Voices of Japanese Feminism*, Berkeley: University of California Press.

*Migration News* (2001) 'Japan: foreigners', February, 8, 2. Online. Available http://migration.ucdavis.edu/mn/archive_mn/feb_2001_13mn.html (16 October 2001).

Ministry of Health and Welfare (1993–2001) *Vital Statistics*, Tokyo: Ōkurashō.

Ministry of Justice (1991–2000) *Annual Report of Statistics on Legal Migrants*, Tokyo: Ōkurashō.

Ministry of Justice Immigration Control Bureau (1998) *Immigration Control*, Tokyo: Ōkurashō.

Ong, A. (1995) 'Women out of China', in R. Behar and D. Gordon (eds) *Women Writing Culture*, Berkeley: University of California Press.

Ortner, S. (1996) *Making Gender: The Politics and Erotics of Culture*, Boston: Beacon Press.

Osteria, T. (1994) *Filipino Female Labour Migration to Japan*, Manila: De La Salle University Press.

Pertierra, R. (ed.) (1992) *Remittances and Returnees*, Quezon City: New Day.

Piper, N. (1997) 'International marriages in Japan', *Gender, Place and Culture* 4, 3: 321–38.

Rafael, V.L. (2000) *White Love and Other Events in Filipino History*, Durham and London: Duke University Press.

Rafael, V.L. and Abraham, I. (1997) 'Introduction', *Sojourn* 12, 2: 145–52.

Samonte, E. (1992) 'The psychological costs of post-employment of overseas workers', *Philippine Journal of Public Administration* 36, 3: 282–94.

Sellek, Y. (1996) 'Female foreign migrant workers in Japan: working for the yen', *Japan Forum* 8, 2: 159–75.

Suzuki, N. (2000a) 'Women imagined, women imaging: re/presentations of Filipinas in Japan since the 1980s', *U.S.–Japan Women's Journal English Supplement* 19: 142–75.

——(2000b) 'Between two shores: transnational projects and Filipina wives in/from Japan', *Women's Studies International Forum* 23, 4: 431–44.

——(2001) ' "Misbehaving victims": reading narratives of "Filipina brides" in Japan since the 1980s', paper presented at the *53rd Annual Meeting of Association for Asian Studies*, 22–25 March, Chicago.

Tadiar, N.X.M. (1997) 'Domestic bodies of the Philippines', *Sojourn* 12, 2: 153–91.

Torrevillas, D. (1996) 'Violence against Filipina OCWs', in R.P. Beltran and G. Rodriguez (eds) *Filipino Women Migrant Workers: At the Crossroads and Beyond Beijing*, Quezon City: Giraffe Books.

Trager, L. (1988) *The City Connection: Migration and Family Interdependence in the Philippines*, Ann Arbor: University of Michigan Press.

Ventura, R. (1992) *Underground in Japan*, London: Jonathan Cape.

Yamashita, A. (1995) 'Kirisutokyō to mizuko kuyō (Christianity and memorial services for aborted foeti)', *Kyōka Kenkyū* 113: 72–84.

Yamazaki, H. (1987) 'Japan imports brides from the Philippines: can isolated farmers buy consolation?' *AMPO* 19, 4: 22–5.

# 7 Resisting history

## Indonesian labour activism in the 1990s and the 'Marsinah' case

*G.G. Weix*

The case began unobtrusively on 24 May 1993 with an item in the Criminality section of *Tempo*, a widely read investigative news magazine published during Indonesia's New Order (1966–98). The Jakarta-based magazine was the first to report the murder of a young female worker and trade union activist from East Java. The story quickly drew the attention of both the Indonesian news media and other news services. Throughout 1993 and 1994 the Legal Aid Society in the nearby city of Surabaya and the Amnesty International Asia Watch teams conducted investigations and circulated their reports on the Internet. The 'Marsinah' case, as it came to be known, publicised the murder investigation, trial and aftermath of a single worker's demise; more broadly it publicised the repressive conditions under which industrial workers renewed trade union activism in Indonesia during the 1990s (Hadiz 1997). The case heightened international scrutiny and drew attention to military intervention in disputes between workers and management in the industrial workplace. For the purposes of this volume, I analyse the politics of representation in the news reporting of this case, and the ways it indexed Indonesian workers as women workers in the broader media and discourse on human rights.

Indonesian and international observers focused on the social identity of the murder victim, a 23-year-old worker named Marsinah. Ironically, the modes of representation – narrative, visual and artistic – which seized upon this young woman's name and face as an emblem of Indonesian workers, concealed as much as they disclosed about her life. This chapter examines the discursive processes which transformed Marsinah from a murder victim into 'a shining symbol for workers' rights' in the Indonesian and international media (Mohamad 1994: 7). I compare the national media's efforts to eulogise her with the local responses of other female factory workers to remember her in East Java. Both national and local responses highlighted gender as an aspect of labour politics; however, they diverged in the ways their narratives portrayed women workers as social actors. Public expressions to honour Marsinah augmented as well as eclipsed the agency of other female industrial workers who, in the aftermath of the publicity, sought to commemorate for themselves a fellow Muslim activist.

Gender was a significant aspect of the initial notoriety of the case. The Indonesian media was persistent in their investigation and relentless repetition of the victim's face and name in headlines and news magazines; they kept the unsolved murder politicised for over three years. The journalists' strategy drew on a familiar mode of Indonesian political discourse: to represent a social category (such as workers, peasants and rural Javanese) with a single name.[1] This narrative device is shared by ethnographers of the industrial workplace who begin their analysis with an abbreviated story of a day in the life of a single worker (Wolf 1992). In the Marsinah case, this narrative strategy reached its grim conclusion when journalists recounted her life story as significant only after, and perhaps because, she was dead.

Because the news media focused on the violent death of a young unmarried female worker, it also marked the emergence of gender and identity politics into the discourse of Indonesian labour politics and human rights debates in the 1990s. The evidence detailing the rape and torture interjected violence against women into public debates in a visceral way. Indonesian trade union movements have confronted violence and state-sponsored repression since their inception in the early part of the twentieth century; male workers have been intimidated through threats of force against organising or engaging in strikes (Ingelson 1986; Hadiz 1997). Yet the Marsinah case resonated with contemporary audiences because the violence against workers could be condensed in the figure of a single woman violated and left for dead. Eulogies appealed to consolidating her status as a worker and as a woman and claimed Marsinah as a representative of both groups (Mohamad 1993a).

National media employed discursive and visual means to represent Marsinah as a victim of state violence; they also reported on local forms of collective action and memory. In East Java, other female factory workers participated in religious and personal forms of memorial such as pilgrimages to her grave (*ziarah*). Their agency can be glimpsed in their everyday discourse and collective action and the ways they remembered Marsinah through personal sacrifice. The national effort to make Marsinah an emblem of the labour movement in the 1990s was laudable, but it inadvertently consigned to all other workers the background role of witnesses. I contend that female factory workers and their local responses to Marsinah's death are as significant as the national campaign to memorialise her. In their local actions we can see the cultural struggles to resist national and international renderings of workers as victims in Indonesian history.

## Official and unofficial versions of the case

The official version of events began with reports in the national newspapers and magazines *Kompas*, *Tempo* and *Editor* which are read by the urban middle class and which represent the best of investigative reporting in Indonesia. On 24 May 1993, the national magazine *Tempo* reported an anonymous murder victim: 'Strike (*aksi mogok*): Who Killed a Worker?' (Himawan and Nugroho

1993: 39). A pen and ink drawing of a figure lying face down among placards depicted the victim as a worker at the firm P.T. Catur Putra Surya, in Sidoarjo village, East Java. The victim had disappeared on 4 May from her rented room just after participating in a demonstration (*unjuk rasa*). Her body was found with signs of severe torture in a rice field culvert in Nganjuk, a smaller village nearby, four days later. Initial reports hinted that the case involved a sex scandal. Throughout 1993, official and unofficial versions and rumours of Marsinah's murder circulated in rapid succession. Earlier versions suggested that her death had been the result of a rape, and downplayed her status as a trade union member engaged in political activity. Later versions speculated on a conspiracy to disguise the murder of a trade unionist with signs that a rape had taken place, and her murder was interpreted as a reprisal by local military for revealing their intervention in a factory strike and dispute. Because identification of the body was not announced to the press until three weeks after it was found, the Legal Aid Society focused on possible collusion between the military and local factory management to hide their involvement from view. Similar to many Indonesian political rumours, the issue of a sex scandal continued to deflect attention towards the personal life of the victim. According to a 22 December 1994 newscast by DeTIK, an Indonesian Internet news service, 'The authorities then appeared to launch a counter-move, suggesting she was killed over a love triangle or an inheritance, even though she did not have a boyfriend' (DeTIK 1994: n.p.).

In August 1993, the Legal Aid Society in Surabaya declared an independent investigation of the factory strike and other events preceding discovery of the body. They reported that Marsinah had joined sixteen workers to negotiate with the factory management for shop agency from the government-sponsored trade union, the All Indonesia Employees Union (Serikat Pekerja Seluruh Indonsia (SPSI)). On 2 May 1993, Marsinah had helped to organise a strike for 500 workers to demand supplemental pay of 550 Rupiah for food and transportation to be added to the daily base pay of 1,700 Rupiah. Their demands stipulated pay increases whether or not women workers took days off for menstrual leave. Twenty-four workers presented these demands to factory management and the management reportedly conceded after two days. Workers, however, voiced a different version to the Legal Aid Society investigative team; they claimed negotiations had yielded only an agreement to mediate the dispute. They reported that sixteen workers were summoned to the local military command (KODIM) on 4 May 1993. Thirteen workers showed up and were forbidden to enter the factory. At the military office, they were forced to admit that they had intimidated the others to strike and to sign resignation letters for 125,000 Rupiah severance pay. Their signatures were witnessed by the government-sponsored union SPSI representative and by two military officers.

Although Marsinah was not one of this group, she planned to take a grievance on their behalf simultaneously to the Departments of Justice and Labour, the police and the SPSI union. Investigations hypothesised that she went to the local military command to ask about her colleagues' dismissals and was picked up later at a coffee house by military officials and taken to be interrogated.

The connections between the factory and the local military command were not coincidental. The military reportedly was linked to the security guards at the factory where Marsinah worked. *Tempo* magazine (Yarmanto and Nugroho 1994: 74) reported that Suwono, a guard at the factory and a former naval officer, 'enjoyed wearing his uniform to intimidate women in the workplace'; the magazine quoted him saying to workers, '[Marsinah] learned her lesson. Now she's dead.'

On 22 March 1994, the Indonesian Legal Aid Society opened a second, independent investigation of the Marsinah murder based on their interviews since August 1993. Doubting both official and unofficial versions of the murder, the investigators hypothesised a more complete chain of events. Their investigative purpose had expanded to include the kidnapping of the murder suspects brought to trial in early 1994. On the inauspicious dates of 30 September and 1 October 1993, police had arrested nine factory employees – managers and a security guard. At the trial in January 1994, the prosecutor had hypothesised that both military and factory officials were involved in Marsinah's abduction and she had been killed primarily because she had threatened to expose the factory's falsification of a watch brand.

> they [the factory management] were reportedly fearful Marsinah would reveal the factory was illegally producing brand-name watches. Accordingly, they arranged to have Marsinah killed. She was picked up by a security guard working for the owner, transferred to another car where she was bound and gagged and taken to Yudi Susanto's [the personnel manager's] house in Surabaya where she was given no food nor water for three days and then killed.
>
> (Human Rights Watch/Asia 1994a: 7)

In addition, the prosecutor had contended that Marsinah had been singled out by the military for violent reprisal and emphasised that she alone had challenged her colleagues' dismissals in writing. The Committee for Solidarity with Marsinah confirmed that she had threatened factory management with a written statement:

> Marsinah had indeed had a meeting with Yudi Astono (CPS manager) at 10:30 p.m. on 4 May. Thirty-three people had showed up and Marsinah had been very vocal, insisting on the raise of 550 rupiah to a total of 2,250 rupiah in accordance with the minimum wage based on the decision by the Minister of Labour, Number 50/92. Marsinah had handed Yudi her 'truf' [trump] card – a letter containing inflammatory knowledge of illegalities at the factory and addressed to the director of the CPS factory.
>
> (Surawijaya 1993: 28–9)

Lane (1994: 2) noted that

> during the period of the trial in January 1994, a mysterious pamphlet appeared entitled 'Marsinah'. It sold nationally as well as at kiosks outside the courtroom. Published by an unknown 'Surabaya Metropolitan Press' and designed to look like an underground publication, it presented a full police version of events.

Unofficial versions of these events were circulated as well. International observers took an interest in the trial in 1994 because the United States Trade Commission was evaluating Indonesia's status as a trading partner. The official and unofficial versions of the events soon became inextricably intertwined as they circulated (in English) on the Internet. Human Rights Watch summarised the combined versions of the trial records:

> Earlier [on] May 4th Marsinah had sent a letter to the factory management protesting the firing of thirteen employees after they went on strike, demanding an increase in wages and threatening to expose the production at the factory of false name brand watches. That evening a security guard at the plant named Suprapto came to the dormitory where Marsinah lived [at] about 9:30 p.m. and invited her out to eat. She got on the back of his motorcycle and was then driven to an intersection where a company-owned car was waiting. ... Marsinah was told she was being taken to the house of the personnel director of CPS, Yudi Susanto ... to discuss a letter she had written protesting the firings. On the way she was bound and gagged. At the house she was tied to a chair, left for three days without food or water, beaten and tortured. On May 7th, the group who had abducted her, led by Ayip, decided to take her to Nganjuk, where she was taken out of the car and killed. An autopsy showed she had bled to death. A sharp instrument had been inserted in her vagina, reportedly on Yudi Susanto's orders so that her death would look like a rape-murder, rather than a result of her labour activism.
>
> (Human Rights Watch/Asia 1994b: 4)

By 1994 Human Rights Watch had expanded its investigation to include monitoring the treatment of the murder suspects as well. The eight CPS executive employees and one guard arrested on 1 October 1993 were detained by military intelligence for three weeks before being moved to the police station and formally charged with the murder (Surawijaya 1993). The Human Rights Watch/Asia's (1994b) *'Openness' Report* condemned the entire proceedings of the arrest and trial of the ten suspects due to evidence of torture during detention. The unlawful interrogations and detentions tainted the record of the facts of the case. Charges against members of the factory management were later overturned by a court of appeals.[2] In June 1995, the Indonesian Supreme Court overturned all convictions of this group.

When *Tempo* magazine (Yarmanto *et al.* 1994: 64) reported euphemistically on the torture of the suspects – that police had given the detainees 'a lesson' (*pelajaran*) – it quoted Yudi Susanto as saying, 'Not enough that I have to confess to killing one Marsinah, I must confess to killing one hundred Marsinahs. What can I do [*sic*]; I will. Who can stand to be tortured continuously like this?' The repetition of her name reverberated in the shared discourse of police and suspects; her death became a series.

Indeed, in the Indonesian national media, more victims of violence began to be described as 'Marsinah'. In reference to Titi Sugiarti, a worker at the P.T. Kahatex Rancaekek firm in Bandung, West Java, who had drowned mysteriously in the Deli River, *Editor* magazine (26 May 1994) referred to her as 'Marsinah [from] Medan'. The caption of a political cartoon in *Tempo* magazine (19 May 1994) reads, 'A Second Marsinah: Titi in the Pool (*Marsinah Kedua: Titi dalam Kolam*)', as if the first death had initiated a series. In the cartoon a man reading a newspaper entitled 'Marsinah: Workers' Heroine (*Marsinah: Pahlawan Buruh*)' frowns, saying, 'The first [case] is still unsolved (*Yang pertama saja masih kusut*)'. In the press, Titi's death was framed also as a sex scandal: the autopsy was cited as stating that she was not a virgin, and the news articles alluded to her 'male guests at the women's dorm (*tamu lelaki di mes wanita*)' (Bakarudin 1994: 56). A workers' committee criticised the investigation:

> West Java police chief says the death of labour activist Titi Sugiarti was accidental drowning, but he would not say it was suicide and called on more witnesses to come forward. A solidarity committee for Titi has pointed to deliberate erasure of evidence at the site of the incident and to poor standards in the autopsy done in Bandung.
>
> (Bakarudin 1994: 56)

Throughout 1994, workers across Indonesia organised trade union actions and protests of military intervention in labour disputes. A third death in Medan was also cited as part of this series. *Editor* magazine (26 May 1994) also referred to this dead worker, Nipah, as 'Marsinah [from] Medan' (cf. Pabotinggi 1994).

## Marsinah's face as an emblem

From its initial status as a murder case to the wider investigation of the trial detainments, the Marsinah case was the most widely publicised news item in 1993 and 1994.[3] *Tempo* magazine featured her on its cover three times (Rahadian 1993: 40). Its annual review issue (8 January 1994) juxtaposed her face with those of two elite women: Megawati, then leader of the Indonesian Democratic Party (PDI) and Syuga, a corporate businesswoman notorious for repressive policies and poor labour conditions in her own factories. Even this triptych image of Indonesian women referred to Marsinah implicitly: 'Not just [the year of] Mega and Syuga (*Bukan Hanya [Tahun] Mega dan Syuga*)'.

Marsinah's name bore an inordinate burden in these efforts to publicise the violent reprisals to trade union actions. Indonesian writers cited the case as the epitome of social protest. The blurred photograph of her face punctuated countless news stories to remind the reading public that the stakes of trade union activism had grown deadly. Public support for improved labour conditions condensed on this image of one single woman. 'The murder galvanized the movement for labor rights as no other incident has in memory. Even the puppet trade union (the government-sponsored SPSI) called her a national hero' (Lane 1994).

The photograph that appeared relentlessly in weekly news magazines was elaborated by artists as well. Realistic sketches gave way to surrealist distortions of her features. In one sketch, a window blind unrolled over her face; in another, one eye was obscured by a swirling montage of colour (Lubis 1993; Nasutiun 1993). On 12 August 1993, the Surabaya Arts Council gathered various artistic renditions of Marsinah for an exhibit to commemorate the hundredth day after her death. The exhibit featured a bronze bust normally reserved for national heroines; it also portrayed murals of events leading up to the murder, similar to the dioramas of historical events in Indonesian national museums. Moeljono, an artist famous for his community art (*seni masyarakat*), was scheduled to attend the opening ceremony. The Surabaya police shut the exhibit down three hours before its opening, claiming that it was a threat to stability and public order with the statement, 'There are elements who are trying to politicize the death of Marsinah through art.' The Arts Council director said it was the first time in twenty-one years that an exhibition had been closed (Human Rights Watch/Asia 1993: 7).

## Marsinah's place in writing

Both journalists and artists portrayed Marsinah as a woman, worker and activist who took ordinary, though courageous, steps to protest her fellow workers' dismissal and rights to organise. Her mode of protest emphasised her role as a literate citizen. She chose to write letters to document production illegalities and to protest the military intervention in the strike. The act of writing distinguished her as a literate worker willing to commit herself to the public record. If she did carry written statements to the factory headquarters or the military command, the letters were lost or destroyed when she disappeared. However, the knowledge that letters may have existed haunted the case.

Marsinah also represented a shift of the social identity of the Indonesian workforce (Jones 1994). A rural villager, she had graduated from junior high school and had taken informal adult courses. She spoke Javanese and Indonesian, and was said to have studied English. Thus, her ethnic, national and international affiliations were a modern hybrid that gave the case added significance. While the media attention to the rape linked her gender with violence, the metatextual status of her multilingualism showed an Indonesian woman worker employing literacy as a new form of agency facing management and military intimidation in labour policy and legal process.

Kathleen Canning (1994: 378) has called the point at which women workers are known for more than their gender the 'linguistic turn' in the analysis of women's labour movements in the twentieth century. Women workers' agency becomes predicated not simply upon collective acts of organised resistance, but upon their status as educated and literate citizens who can mobilise petitions and letter campaigns in response to state violence. Although Marsinah's photographic image and name were repeated to become a series, her singular act of writing became the most enduring and significant mode of the politics of representation and publicity about the case. In one of the two editorials (Mohamad 1993a, 1993b) he wrote to maintain the visibility of the murder case and its investigation, Goenowan Mohamad (1993b: 77), then the editor of *Tempo* magazine, commented:

> What is Marsinah to me, to you? She is a name in the news, and between us and her there is a gap, or distance. News tells us what happened, or is happening, a presentation [of events] already defined and separate from us. The newspaper, magazine, radio or television can cross time and space, but at the same time they cannot bring us closer. Marsinah's death becomes a *news item* [*sic*, English], among a pile of news items, and tomorrow it won't be there. Maybe the death makes us angry, horrified, questioning, but it is ephemeral ... Marsinah is not present. She is only represented by a name ... Marsinah: at last we know that the woman who met her death in a warehouse isn't going to appear. She is only represented by her name. And aware or unaware [of her], we feel we don't need an appearance (*kehadiran*). We are continually connected [to her] with a headline.

Goenawan Mohamad describes how, for a reading public, a death becomes a 'news item' (in English, no less). Collective awareness in the modern era relies on 'headlines' rather than 'appearances' that in Javanese social contexts would signal ghosts. The role of language in news media bypasses other forms of cultural memory:

> post-modernist theory might, according to Derrida, posit language as a structuring reality, a partial explanation as to why an event, even a murder, can be assimilated into a series of signs and linguistic/semantic distinctions (which constitute readers' experience of history as a flow of events from which one can choose to attend to some events, and not to others). Language is a process which does not halt our attention; we must intervene to mark certain events, and in this case, a particular death.
>
> (Mohamad 1993b: 77)

Although reading the daily news can create an amnesia for specific events, Mohamad argues that we should arrest our attention on those stories for which emotional response is inadequate. The seamless act of reading so many 'news items' must be interrupted by the ethical imperative of apprehending this one case. Mohamad continues:

Maybe there is nothing outside the text, as Derrida says. Maybe as in the film *Rashomon*, the perspective of the victim is only one among many. But facing those who are tortured and killed for no reason ... we cannot talk simply of determining an exegesis of reality, or a point of view.

(Mohamad 1993b: 77)

This particular death compelled Goenawan Mohamad to recall her name, face and violent death to arrest the amnesia of a reading public. In his second editorial in December 1993 (later published in English in 1994),[4] he recounts the details of Marsinah's life with the familiar device of narrating the life of a single individual:

[She was] an adopted child who worked selling snacks and who did not continue school beyond high school. She attended computer and English classes. To learn more, she read newspapers and watched television at a neighbor's place. She once told an acquaintance, 'Knowledge will change one's destiny.'

(Mohamad 1993a: 34)

Goenawan's sketch of Marsinah's life culminates in her career as a labour activist. In his speech at a ceremony honouring her contribution to workers' struggles he claimed that Marsinah represented both workers (*kaum buruh*) and women (*kaum perempuan*) because she died through the rape and the humiliation of torture to a female body (Mohamad 1993a). His eulogy identified the violence of rape as particular to women's experience; therefore her death fused class and gender interests. Thus Marsinah's name becomes both adjective and scandal for 'women' in Indonesian media and national discourse in the 1990s. The result of the Marsinah case was to publicise all women workers' interests through the images of violated femininity, and to evoke the imperatives of workers' rights predicated on the documentation of human rights abuses of torture and rape. Journalists argued that the case gave 'voice to workers' (Human Rights Watch/Asia 1994a) just as Goenawan Mohamad's eulogy (Mohamad 1993a: 34) enshrined Marsinah with her own metaphorical voice for political action:

Marsinah is a shining symbol of the fight for human rights. ... The bosses and security apparatus took action to silence the rebellious workers. Marsinah was the loudest and most upsetting of all, and for that she was murdered ... Marsinah was victimized both because she was an activist and because she was a woman. Rather than be ignored, the death of this village woman drew international attention ... they went on strike to demand better wages, and could not be ignored. Thus, women have ceased to be marginal.

## Local responses to her death

Local responses to Marsinah's death were also reported in the national and international media. In 1994, while journalists were eulogising her in Jakarta, a *Tempo* reporter described her grave in Nglundo village, Nganjuk, East Java, as 'crowded with workers who had come from several cities on pilgrimage (*ramai diziarahi para buruh yang datang dari berbagai kota*)'. East Javanese workers had arranged a day of mourning for her. The reporter noted that since 1993 their wages had risen to 3,000 Rupiah a day (Hakim and Widjajanto 1994: 40). Still, it was no small sacrifice for the young women to sew themselves black dresses for the journey since each dress cost 5,000 Rupiah, and the round-trip bus ride cost 7,000 Rupiah. One of Marsinah's friends, Sri Widarti, 23 years old, was quoted as saying, 'Marsinah is a model for workers (*Marsinah adalah teladan bagi buruh*)' (Kustiati *et al.* 1995: 87). Photos of the event show young women in full veil and holding a flower wreath with the caption 'Marsinah, your struggle is our bridge (*Marsinah: Perjuanganmu adalah jembatan kami*)' (ARM[5] *et al.* 1994: 73). The workers' wreath addresses Marsinah, not the reading public, with the pronoun '*kami*', meaning 'we, not you'. In Indonesian political rhetoric '*kami*' signals a collective identification and invitation to join the group.

In contrast with other renditions of her life circulated on the Internet in 1994, these workers did not attempt to speak for Marsinah. Rather, East Javanese workers engaged alternate modes of collective action to remember Marsinah and keep her present. Their actions and words exemplify how some Javanese experience the dead – even those who are violently murdered – as still connected to the living. For example, one journalist observed a young woman approach the grave with flowers and the greeting, 'Good morning, Marsinah (*Selamat pagi Marsinah*)' (Passe 1993). These young women attempted to keep Marsinah singular and present rather than relegating her to the past. Their collective actions provided alternate forms of social memory to a literate public accustomed to reading 'news items'.

The young women also drew attention to themselves as a social group, 'us (*kami*)': that is, unmarried female workers similar to Marsinah. Sociologists note a distinct pattern of industrial employment for young women in Java; they work before marriage and the birth of their first child (Grijns 1994). This workforce has been unusually subject to modes of managerial control (Wolf 1992; Ong 1991). In the broader context of Southeast Asian industrialism, Maila Stivens (1994: 375) argues that 'studies of women as gendered subjects have focused on male control and exploitation of women's labor in newly industrialized factories'. However, she criticises James Scott (1985) for his neglect of gender in his analysis of agency for Malay peasants since his own ethnography reports that women organised boycotts. Stivens concludes that 'gender relations are not just an "effect" of the continuously evolving social transformations set in train by imperialism and colonialism, but are intimately involved in producing those transformations' (Stivens 1994: 376).

As a facet of social identity, gender is predicated not only on kinship and social reproduction, but on collective action as well. Although young unmarried women (*gadis*) are not considered social adults in Indonesian or Javanese society. Unlike students (*siswa*) or youth (*pemuda*), their actions are not expected to transform society with a social and moral force (Brenner 1996). Young female workers became significant in the national media and arts because the discursive process to universalise Marsinah as an icon of protest drew attention to their collective status as young unmarried female workers. Suruchi Thapar (1993: 82–3) has discussed a similar process at work in India's nationalist movement. If gender relations in some sense 'produce' the social transformations accompanying industrialism in Indonesia, as Stivens argues, then the specific practices of these young women workers are themselves transformative. When they act collectively to remember Marsinah through public pilgrimage, they draw attention to femininity, youth and other distinctive features that define their social status outside the workplace.

Aihwa Ong (1991: 306), whose analysis of factory workers in Malaysia is well known, argues for apprehending 'the lived realities of women on the new frontiers of industrial labour, workers' actual experiences and the creativity of workers' responses in the late 20th century'. While a feminised working-class solidarity has been anticipated by numerous scholars, Ong (1991: 306) warns us that:

> workers' struggles and resistances are often not based upon class interests or class solidarity, but can instead comprise individual, and even covert acts against various forms of control. Workers wage 'cultural struggle', that is, battles over cultural authenticity among different ethnic and class groups in a nation state … Interests are linked to kinship and gender more than to class. Workers struggle over cultural meanings, values and goals.

The East Java workers' reported attempts to remember Marsinah's identity and to memorialise her death are one such struggle. The politics of representation surrounding this case traverse national and international arenas, and do not always coincide. National journalism as well as legal and artistic depictions eulogised her as a martyr. Local workers speak to her at the gravesite as if she were still present.

Cultural meanings also devolved upon Marsinah's body after her death. Ong (1991: 307) writes, 'Ultimately, the body is the site at which all strategies of control and resistance are registered. Bodies are disassembled and reassembled in commodified images of essentialized women who work.' Journalists and other observers focused on the rape and torture of a violated female body, which became a sign of defeat for labour activism. In the aftermath of investigation, state officials and critics shared an almost obsessive concern for repeated autopsies to determine the manner of death. In August 1995, a second exhumation was sought for a fourth, independent round of forensic and DNA analysis (Yarmanto and Nugroho 1994).[6] Each additional autopsy was justified by the search for more scientific knowledge and verification; yet they also mark this death as a source of anxiety. Thus, the Marsinah case indexed the body of workers as female and an origin for knowledge about workers-as-women in the Southeast Asian industrial transformation.

As part of these struggles over the cultural meaning of Marsinah's life and death, the young women workers in East Java who go on pilgrimage to her gravesite in Nglundo emphasise it as an honoured place. Their references to Marsinah's actions as a 'model' and 'bridge' do not evoke a feminised identity (or death). They do not dwell on images of violence done to a female body. Instead, their collective acts of commemoration strive for a politics of social memory in which they are actors as well; gender is implicit in their collective actions of remembrance and personal acts of sacrifice. The national discourse reports about them, but rarely acknowledges the ways their actions diverge from the symbolism of victimisation.

## Conclusion

Marsinah has been honoured with three posthumous awards (Kusumah 1993; n.a. 1995). To dedicate awards to her underscores the discursive strategy of national media and opposition groups to testify to the violence which suppressed her activism. Journalists say they feel haunted by the case.

> Marsinah who was killed so horribly two years ago still cannot leave our hearts at peace. Why is our society so *concerned* [*sic*, in English] with Marsinah? Because she was killed at the moment the country was affirming to the world its commitment to respect human rights.
>
> (Maulani 1995: 29)

To honour Marsinah, or to register her as an unquiet spirit, can obscure the effects of her political activism and the political agency of other young women workers. Indeed, the more her case was publicised, the more the details of torture and rape eclipsed her original concerns to improve labour conditions by increasing wages, gaining shop agency and securing the right to organise free of military intimidation. Human Rights Watch/Asia (1994a: 8–10) noted:

> Indonesian workers lack the basic right to freedom of association ... Labor rights are a sensitive issue for three reasons: first, economic growth and foreign investment are dependent on cheap labor costs; second, labor activism is associated with the banned communist party, and third, [the Marsinah case] was an embarrassment to the APEC summit in 1994. The threat of the U.S. to suspend GSP status [was] because of [these] violations of workers rights.

In both internal and international discourse about Indonesia during the 1990s, Marsinah's death signified a new concern over human rights abuses. As Mulya Lubis (1993: 26) wrote in an editorial headlined 'Marsinah': 'There are those who think it is no longer important who killed Marsinah: army, police, security officers or thugs. It's important to take inspiration from her spirit to support workers' rights.' That rhetoric unfolded in narratives of her life's story as a worker, woman and activist that culminated in the reprise of a violent death. The political effects of the reportage were to sustain pressure on the Indonesian state to revoke military intervention in labour actions and to improve minimum wage laws; in the cultural arena, the case became a media legend about gender and violence.

The killing of Marsinah resonated in Indonesia more than any human rights case in recent memory. The fact that she was young and female was part of it; the fact that she was so brutally murdered was another. Her murder came at a time of unprecedented domestic and international scrutiny of Indonesia's labour rights practices, and Marsinah's efforts to intercede on behalf of thirteen workers sacked and detained by the local military gave her, after death, something akin to sainthood (Human Rights Watch/Asia 1994a).

The publicity that reduced Marsinah's name to an adjective for scandal produced some structural reforms, but critics argued that the reforms simply removed intimidation further from view. Abdul Latief, then Minister of Labour, revoked the regulations on 16 January 1994 (Human Rights Watch/Asia, 1994a: 1) that permit military intervention into industrial disputes. These regulations were legitimated in 1986 and passed by Admiral Sudomo, Indonesia's first Minister of Manpower in the era of Indonesia's rapid industrialisation (1985–90); the regulations allowed military intervention in cases pertaining to strikes, work contracts, dismissals and changes in status or ownership of a company. However:

> Teten Masduki, the coordinator of Worker's Solidarity Forum, said the withdrawal of the decree simply removed military intervention from international scrutiny and view altogether. Security forces are governed by their own set of 'autonomous' regulations which allow them to intervene in industrial disputes if they pose a threat to national security or stability. SPSI is dominated by military and security personnel, and many military have business interests. Intervention, as legitimized by these decrees, has clearly not been effective in containing the problem, and the regime is well aware of this. If they choose to persist with this policy, their methods will become more violent. *There will be more victims, more Marsinahs.*
>
> (Jana 1994: 1, my italics)

Saraswati Sunindyo (1995) has analysed the murder-as-event in the Indonesian New Order media discourse as a process by which murders of women are reported as sex scandals to shield the male public officials involved from view. This case began to unfold in that logic. Gradually the event-as-murder revealed Marsinah as a woman worker and trade union activist. At its margins, journalists and activists noted alternate modes of political representation and memory (Goodwin and Bronfen 1993: 3). Ironically, the publicity of this case symbolised workers-as-women in order to forward their claims to justice, but did not claim their own local politics of memory as valid forms of agency. Workers' actions inaugurated themselves as visible social actors to a national reading audience, but with different symbols, social positions and collective acts.

Marsinah and her unquiet death evoked local and global responses that were as distinct as they were intertwined. In 1996, groups of local workers, non-governmental organisations and other young women workers held a thousand-day ceremony for Marsinah and went again on pilgrimage to her grave (*Republika Online* 1996). National media emphasised her singular and feminised violent death as a scandal, and relied on a barrage of images to keep the story alive as a 'news item'. Visitors to her grave engaged her in conversation. Both efforts kept Marsinah 'alive' but in very different senses. Avery Gordon (1997: 26) has commented on the role of ghosts in sociological discourse: 'The ghost is not simply a dead or a missing person, but a social figure, and investigating it can lead to that dense site where history and subjectivity make social life.' Those young female workers who participated in Islamic rituals marked her passing as a fellow Javanese Muslim. They undertook extraordinary personal sacrifice to visit the place in the East Javanese landscape where she is buried. Their actions and words resist concluding Marsinah's life and death as merely testimony to state repression and violence. Instead, her grave becomes a place where Indonesians can speak to her and draw close to an ongoing struggle for cultural meanings of labour politics in history.

## Postscript: December 1999

The last two years have catalysed not only unimagined changes in Indonesia's national presidency and politics, including the long-awaited referendum and independence of East Timor, they have shown also that even such dramatic publicity in one sphere of politics does not necessarily lead to a significant rise in Indonesia's international profile on issues of trade and labour rights. Only this past month as demonstrators clashed in Seattle in protest of the World Trade Organisation, the cases of labour rights abuses were generalised across much of Asia. Certainly, cases such as the Marsinah murder initiated the link-ages between local struggles by women workers and international policy and its critics. However, as I have argued above, the shape and texture of those local struggles, their rhetoric and form, often remain obscure to many international observers who are sympathetic to the rights of Indonesian workers – both male and female – and who also take civic and political action to protest the restrictions enforced by national laws and informal practice. Whether that obscurity is due to a dearth of translation or simply the world's continued ignorance of Indonesia is difficult to say. Since 1998 it has been more difficult to ignore Indonesia, although much of the publicity is increasingly crossing the linguistic divide of 'global English'.

It is perhaps telling that during the time this article has been written, revised and in press, women in Indonesia have moved into new and more visible political forums as new political party members, yet still declaring their aspirations and inspirations in the idioms of their gender and religious affiliation as Muslims, as possibilities as yet unrealised. When Ibu Hamama, the current vice-chair of Aisiyiyah, the national women's organisation affiliated with Nahdatul Ulama, was asked if a woman could be president, she did not answer directly, nor did she mention (then) contender Megawati Sukarnoputri. Instead, she couched her response in the Indonesian context: 'Yes, Islam allows that' (interview 1998 by Jennifer Bright, doctoral candidate at the School of Oriental and African Studies, reported at the Association of Asian Studies meetings in March 1999). Just as Marsinah's supporters reflect on their connection to a singular woman made famous by unfortunate and violent acts, so do Indonesian women leaders now reflect on the irrevocable changes of this decade which began with the publicity seeking justice for the murder of a trade union activist, and ended with the election and selection of a woman vice-president.

## Acknowledgements

This chapter was originally prepared for the panel *Concealment, Disclosure, and Social Wounds* at the American Anthropological Association meeting, 15–19 November 1995, and presented at the Montana Academy of Sciences and the Women's Studies Research Forum at the University of Montana in April 1996. I thank Victor Montejo, Janet Finn, Anya Jabour, Sara Hayden, Debra Slicer, Barbara Andrew and especially Julie Shackford-Bradley for comments on earlier drafts, as well as the editors of this volume for their assistance in revisions. I should also like to acknowledge Ratna Saptari's forthcoming article, 'The politics of domination and protest in Indonesia: "Marsinah" and its aftermath', and thank Ratna Saptari for comments on this chapter. All translations used in this article, apart from those extracted from the Human Rights Watch/Asia *'Openness' Reports*, are my own.

## Notes

1   Siegel (1998: 20–2) discusses President Sukarno's autobiography in which he tells the story of Sarinah, the nursemaid who inspired his term 'Marhaen', the name of an ordinary peasant farmer for the People (*rakyat*) in his political speeches.

2   The cover of *Tempo* magazine on 19 March 1994 read 'Marsinah: Justice Gone Astray? (*Marsinah: Peradilan yang Sesat?*)'. Three other articles that spring (n.a. 1994; Margono *et al.* 1994; Safrita 1994) also indicated the convictions were suspect.

3   *Tempo*, for example, ran stories on various aspects of the Marsinah case on 30 October 1993 (cover story on '*Menyingkap Kasus Marsinah* (Revealing the Marsinah Case)'); 6 November 1993 ('*Marsinah dan Mutiari Setelah Rekonstruksi* (Marsinah and Mutiari after Reconstruction)'); 18 December 1993 ('*Marsinah dan Politik Perburuhan* (Marsinah and Labor Politics)'); and 8 January 1994 ('*Antara Nipah dan Marsinah* (Between Nepah and Marsinah)').

4   This editorial was a speech made at the Yap Thiam Hien Award presentation. It was subsequently published in English in *Tempo* on 13 January 1994.

5   'ARM' is the acronym used by this particular journalist as a pen-name.

6   The first autopsy was performed on 9 May 1993, the second on 30 October 1993, and the third the following day, 1 November 1993 (Yarmanto and Nugroho 1994).

# References

n.a. (1994) 'Kesaksian dari kamar pembantu (Witness from the servant's quarters)', *Tempo* 2 April: 91–2.

——(1995) 'Marsinah is recommended as a young patriot', *Merdeka*, Jakarta, 16 May.

ARM, Hakim, J. and Widjajanto (1994) 'Mengapa tulang pinggul hancur? (Why was the hip bone broken?)' *Tempo* 21 May: 73.

Bakarudin (1994) 'Antara Titi dan Marsinah (Between Titi and Marsinah)', *Editor* 36, 2 June: 56.

Brenner, S. (1996) 'Reconstructing self and society: Javanese Muslim women and "the veil"', *American Ethnologist* 23: 673–97.

Canning, K. (1994) 'Feminist history after the linguistic turn: historicizing discourse and experience', *Signs* 19, 2: 378–404.

DeTIK (a news service circulated on the Internet) (1994) 'Confused murder scenario'. Online. Available http://www.detik.com.

*Editor* (weekly magazine), Jakarta, various issues.

Goodwin, S. and Bronfen, E. (1993) 'Introduction', in E. Bronfen, E. Bronfen and S.W. Goodman (eds) *Death and Representation*, Baltimore: Johns Hopkins University Press.

Gordon, A. (1997) *Ghostly Matters: Haunting and the Sociological Imagination*, Minneapolis: University of Minnesota Press.

Grijns, M. (ed.) (1994) *Different Women, Different Work: Gender and Industrialization in Indonesia*, Aldershot: Avebury Press.

Hadiz, V.R. (1997) *Workers and the State in New Order Indonesia*, New York: Routledge.

Hakim, J. and Widjajanto (1994) 'Hitam-hitam untuk Marsinah (Wearing black for Marsinah)', *Tempo* 14 May: 40.

Himawan, I.Q. and Nugroho, K.M. (1993) 'Siapa membunuh buruh? (Who killed a worker?)', *Tempo* 5 June: 39.

Human Rights Watch/Asia (1993) *Indonesia: More Restrictions on Workers*, September, New York: Human Rights Watch.

——(1994a) *'Openness' Report: Part IV. Labor Rights and the Marsinah Case*, September, New York: Human Rights Watch.

——(1994b) *Indonesia: New Developments on Labor Rights*, January, New York: Human Rights Watch.

——(1994c) *'Openness' Report: Part IX*, September, New York: Human Rights Watch.

——(1995) *Petition on Indonesian Workers Rights to the U.S. Trade Representative*, June, New York: Human Rights Watch.

Ingelson, J. (1986) *In Search of Justice: Workers and Unions in Colonial Java: 1908–1926*, New York: Oxford University Press.

Jana, D.K. (1994) 'Indonesia repeals labor decree', *Green Left* 128. Available online http://www.igc.greenleft.news (26 January 1994).

Jones, G. (1994) 'Labor force and education', in H. Hill (ed.) *Indonesia's New Order: The Dynamics of Socio-economic Transformation*, Honolulu: University of Hawaii Press.

Kustiati, R., Hakim, J. and Toha, M. (1995) 'Tim khusus untuk pembunuh Marsinah (Special team for Marsinah's killer)', *Forum Keadilan* VI, 3: 86–8.

Kusumah, M. (1993) 'Nasib buruh dan politik kekerasan (The fate of workers and the politics of violence)', *Tempo* 30 October: 30.

Lane, M. (1994) 'Cover-up exposed in Indonesian murder', *Green Left* 138. Online. Available http://www.igc.greenleft.news (13 April 1994).

Lubis, T.M. (1993) 'Marsinah', *Tempo* 30 October: 26.

Margono, A., Kukuh, A. and Hakim, J. (1994) 'Ada rekayasa di balik kematian Marsinah? (Were there conspiracies behind the death of Marsinah?)', *Tempo* 26 February: 71.

Maulani, Z.A. (1995) 'Marsinah' (editorial), *Gatraed* 20 May: 29.

Mohamad, G. (1993a) 'Marsinah', *Tempo* 18 December: 34.

——(1993b) 'Marsinah', *Tempo* 21 August: 77.

——(1994) 'In rural Java, death comes to a fighter and a dreamer', *International Herald Tribune*, Paris, 13 January.

Nasutiun, A. (1993) 'Marsinah dan politik perburuhan (Marsinah and labor politics)', *Tempo* 18 December: 36.

Ong, A. (1991) 'The gender and labor politics of postmodernity', *Annual Review of Anthropology* 20: 279–309.

Pabotinggi, M. (1994) 'Antara Nipah dan Marsinah (Between Nipah and Marsinah)', *Tempo* 8 January: 90.

Passe, Z. (1993) 'Pengorbanan dadis dari Nglundo (The sacrificial virgin from Nglundo)', *Tempo* 30 October: 21.

Rahadian, A. (1993) 'Ini Tahun Marsinah (This is the Year of Marsinah)', *Tempo* 11 December: 40.

*Republika Online* (daily newspaper) (1996) 'Marsinah: 1,000 days, 1,000 riddles', Online. Available http://www.republika.co.id (12 February 1996).

Safrita, A.H. (1994) 'Dicari mayat Marsinah asli (Marsinah's corpse sought)', *Editor* 35, 26 May: 25.

Saptari, Ratna (2002) 'The politics of domination and protest in Indonesia: "Marsinah" and its aftermath', in A. N. Das and M. van der Linden (eds) *Work and Social Change in Asia: Essays in Honour of Jan Breman*, New Delhi: Manohar Press.

Scott, J. (1985) *Weapons of the Weak: Everyday Forms of Peasant Resistance*, New Haven: Yale University Press.

Siegel, J. (1998) *A New Criminal Type in Jakarta: Counter-revolution Today*, Durham: Duke University Press.

Stivens, M. (1994) 'Gender at the margins: paradigms and peasantries in rural Malaysia', *Women's Studies International Forum* 17, 4: 373–90.

Sunindyo, S. (1995) 'Murder, gender and the media: sexualizing politics and violence', in L. Sears (ed.) *Fantasizing the Feminine in Indonesia*, Durham: Duke University Press.

Supartono, Alex (1999) *Marsinah. Campur Tangan Militer dan Politik Perburuhan (Marsinah. Military Involvement and Labor Politics)*, Jakarta: Legal Aid Society Foundation.

Surawijaya, B. (1993) 'Sembilan tersangka untuk satu yang mati (Nine suspected for one dead)', *Tempo* 11 October: 28–9.

Thapar, S. (1993) 'Women as activists, women as symbols: a study of the Indian nationalist movement', *Feminist Review* 44: 81–96.

Wolf, D. (1992) *Factory Daughters: Gender, Household Dynamics and Rural Industrialization in Java*, Berkeley: University of California Press.

Yarmanto, W. and Nugroho, K.M. (1994) 'Buntut kematian Marsinah di Sidoarjo (The final chapter on Marsinah's death in Sidoarjo)', *Tempo* 6 October: 74.

Yarmanto, W., Hakim, J. and Widjajanto (1994) 'Marsinah telanjur tewas (Marsinah irretrievably slain)', *Tempo* 5 March: 64.

# 8 Contradictory identities and political choices

'Women in Agriculture' in Australia

*Ruth Panelli*[1]

## Introduction

*Gwen*[2] *completed her education by studying domestic science: 'I went to a domestic art school, of all things. They taught me to make an apple turnover, wash a man's sock and to thoroughly go through a very strict routine for washing floors. That was going to set me up for life.' She married and set up home on a dairy farm with her husband. Following his death many years later she took over the farm and encountered the full impact of gender assumptions in her industry: 'The agents would come to the door and ask for "the boss". They didn't expect me to be calling the shots.' Throughout her farming life Gwen became increasingly aware of the gendered nature of agriculture. Thus in the 1980s she became involved with local women's farm support and skills courses. These activities led to the establishment of the First Victorian Women on Farms Gathering which has since grown as a model of annual women's conferences held for farm (and other rural) women in many states across Australia. Gwen has attended a number of these Gatherings and subsequently participated in the national group Australian Women in Agriculture (AWIA) and the inaugural Women in Agriculture International Conference.*

Gwen's story is one of a woman who has become part of the recent Women in Agriculture movement. In measure she represents the type of farmer and activist about which this chapter is written. Gwen's story is included here, not because she is a key 'leader' of the Women in Agriculture movement, but because she represents the type of farmer who, through being a woman, has faced certain social expectations and industrial constraints in Australian agriculture. In exploring these situations through the 1990s, Gwen joined thousands of other Australian farm women in recognising her marginal position in agriculture. Together these women participated in a collective farm politics that has sought greater recognition and participation of women across Australian agricultural industries.

The Women in Agriculture movement involves a constellation of informally connected groups, organisations and networks all of which share an interest in the greater recognition and involvement of women in Australian agriculture (Liepins 1998a). Collectively, the movement illustrates a 'politics of locality' – one which is based on women's negotiations of their location and marginalisation in both the farming and industry arenas. Furthermore, the movement has become an alter-native political space in which women can explore contradictory identities and individualised and collective political strategies (Fincher and Panelli 2001). This chapter traces the agency mobilised by Australian women in recognising the contexts and contradictory identities which shape their lives. It highlights how these contradictions have provided political choices from which women have been able to develop a multi-faceted agency and agitate for greater recognition of women in agriculture. After the following brief sketch of the movement, a review of this emerging agency is given. Then, an analysis is made of the strategies women have selected in mobilising collectively around their shared goals. Finally, the chapter explores the scope of this politics by identifying the spaces and impact of the movement across a variety of scales.

The material incorporated in this chapter is drawn from a three-year study of Australian farm women's politics. Data was gathered on the contemporary industrial, social and political contexts from which the movement developed. This was then followed by an intensive period of fieldwork (April 1993 to November 1994) in which I worked as a participant and researcher with three groups within the Women in Agriculture movement.[3] Regionally, I worked with, and studied, the Fifth Victorian Women on Farms Gathering held at Glenormiston in February 1994. Nationally I participated in, and studied, the AWIA, and internationally I was a working (and researching) committee member of the First Women in Agriculture International Conference held in Melbourne in July 1994. The selection of these groups illustrates the impor-tance of activities originating out of Victoria in the early years of the movement.[4] It also illustrates the different types of networks and organisations that could all be identified within this 'polymorphous and polycephaleous' new social movement (Crook *et al.* 1992; Liepins 1996a). The Gathering commenced in Victoria in 1990 and is an annual event 'owned by the rural women of Victoria and … handed on from group to group in trust' (Rural Women's Development Group 1994a: n.p.). Local organisations based in different rural towns take on the commitment to stage the Gathering once a year, providing a weekend programme for women to celebrate their associa-tion with agriculture, share stories, attend keynote addresses, participate in workshops and seminars on agricultural and/or personal development topics. In contrast, AWIA was selected as an ongoing formally incorporated lobby group, and the International Conference was selected as a single-event case group structured around a temporarily incorporated committee formed to stage this inaugural conference of 850 agricultural women coming from thirty-three different countries (Liepins 1996a). A wide range of women were involved with these groups. Although data on ethnicity and socio-economic class was not available, anecdotal evidence showed that women from medium- and large-scale enterprises were participating in the movement and partici-pants were less often from aboriginal or other non-English-speaking backgrounds.

# Developing agency

*Brenda married a farmer in western Victoria. Throughout her marriage she raised her children, was active in a voluntary capacity with a number of local community groups and worked alongside her husband producing wool and grain from their medium-sized property. Following the death of her husband, Brenda continued with the farm until her son and his wife took over the business. Throughout this period, Brenda experienced a range of situations which enabled her to reflect on the social construction of farming and her own multiple identities. Many of these reflections were articulated through the meetings and media campaigns of the Women in Agriculture movement with which she became actively involved. She recognised that her identity was complex: 'I operate in a whole lot of different boxes.' She also identified with other partici-pants in the movement as a 'farmer'. Her experience and frustration at not being readily recognised as a 'farmer' provided the stimulus for a vigorous agency which she claimed and used throughout the 1990s:*

> *I decided somewhere along the line because of what I was encountering as a woman farmer that my daughters were not going to have to put up with some of the shit I'm putting up with now. I want to see women educated to their own worth. I want them to understand how valuable their contribution is ... Women in Agriculture can do this.*

The political activism of Women in Agriculture has emerged from both an indi-vidual and collective practice of agency. This capacity for activism is not a homogenous or universal quality common to all farm women in Australia, but has developed for some women in particular locations and through specific circumstances. Indeed, agency is always occurring 'in place'. It is a contextu-alised capacity.

The context of activism is stressed by Mohanty (1995: 69) who has argued that feminist politics requires an historically and locationally contingent sense of political agency. In particular she has contended: 'my location forces and enables specific modes of reading and knowing the dominant' (Mohanty 1995: 82). In the case of Women in Agriculture, participants and leaders of the movement have developed specific readings and sensitivities to their posi-tions within agriculture and rural society. They have also come to 'read and know' the dominant discourses of masculinity and market-oriented produc-tion that shape much of their experience in agriculture. Participants (like Gwen and Brenda) used the movement as a meeting place to talk about the social expectations placed upon them as 'farmer's wives' and 'community workers'. They talked of how these expectations affected beliefs about how they should act in their local area. So, too, the movement provided an envi-ronment in which the popular and industrial discourses surrounding agriculture could be highlighted and challenged for their overly masculinist character (Liepins 1996b, 1998b).

Women in Agriculture established a strategic agency formed out of the contradictions they identified for women within farming in Australia (Liepins

1998c). Using a post-structuralist notion of agency proposed by writers such as Weedon (1987) and Davies (1991), it is possible to discern how Women in Agriculture participants have recognised the variety of identities and positions in farming (Liepins 1998a). This has included highlighting the many 'hats' they wear, or the 'different boxes' out of which they operate. Cumulatively, events, texts and interviews across the movement revealed numerous common identities recognised by women including: farmer, helping hand, business partner, information manager, mother, community volunteer and off-farm worker.

Women used and described these identity labels in the Gatherings. By sharing these multiply constituted identities, along with the different perspectives they provide on agriculture, activists have recognised how some positions or identities (for example, 'farmers' wives') have marginalised their recognition and participation in agriculture. Moreover, they have noted both the physical spaces and social locations in which these different positions are articulated. Maureen explained:

> It's a real pleasure to be introduced as a farmer, I can tell you, because I have not got the credibility of being a farmer because I'm not doing the hands-on stuff. I don't drive the tractor or do any of the physical stuff and I work off-farm to earn the money to keep it going. But I can manage it, and hell, I am good at the financial bit of it. I know I'm good with the bank managers and with the financial institutions … [While he was alive] I could be part of the farm with my husband but I couldn't be anywhere else and when I went out I couldn't join in farm discussions and it was really frustrating … The gossip session at the post office and at the pub and in the footy shed where men find out a whole lot of stuff about farming, like who is able to buy the peas, what price he can get, all that stuff, as women we haven't got access … I can go up to the post office where all the male farmers are already talking, and they just change the subject.
>
> (Women on Farms Gathering 1990: 14–15)

By identifying the contradictions and power relations involved in being multiply positioned in different ways, Women in Agriculture activists have explicitly engaged with the politics of their locations and made strategic identity choices. Weedon (1987: 125) argues that such identity formation and agency arises from options available to women as subjects of contrasting discourses:

> The subject … exists as a thinking feeling subject and social agent, capable of resistance and innovations produced out of the clash between contradictory subject positions and practices. She is also a subject able to reflect upon the discursive relations which constitute her and the society in which she lives, and able to choose from the options available.

This type of discursive reflection and selection of positions has been apparent throughout the groups in the movement both in their speeches and written texts. For instance, the leader of the Fifth Victorian Women on Farms Gathering explained: 'In preparing this Gathering, it was the intention of the committee to present serious information for professional farming women' (Rural Women's Development Group, 1994b: 47). Likewise, in a most widely reported media campaign, the Women in Agriculture International Conference contended, 'Women in agriculture are not farmers' wives. They are farmers. We need to tackle the issue of our identity to get the recognition we deserve' (Kyne 1994: 33).

While these examples illustrate the strategic promotion of particular identities at specific times of action, it is also important to note that the Women in Agriculture movement was a space where women could negotiate, own and work outwards from a simultaneous location in several different identity positions. Mouffe (1995: 323) refers to this type of experience as the 'articulation of an ensemble of subject positions'. In the case of Women in Agriculture, participants were able to communicate with each other about how they simultaneously acted with, and alongside, the movement: 'Like everybody, I wear lots of hats … in paid employment, on the farm, answering the phone part-time, driving children around part-time; and I'm very involved in a lot of local community groups' (Pam, January 1994).

Cumulatively, awareness of these 'ensembles' enabled women to acknowledge a range of perspectives which aided their political agendas and strategies. For instance, one leader was to explain to me in interviews that she felt able to draw on a whole variety of experiences when campaigning for the movement in either parliamentary or industry circles. This involved her using experiences as a farmer, a mother, the acting farm accountant, the secretary of a local farming organisation, the member of an industry co-operative and the committee member of a regional environmental body. She explained: 'You have to speak their language, but you can use your other experiences to unsettle them too. I mean, they need to know you can balance the books, but you also have other experiences they haven't even thought of' (Sonja, September 1993). In this way, the activist was able to establish an occupational legitimacy (being a 'farmer' managing a business unit) while also lobbying agricultural service industries for funds to support Women in Agriculture. She recounted (and I accompanied her and observed) how in seeking funds from industry organisations she would demonstrate her (and argue, by example, other women's) multiple and complex contributions to agriculture in terms of running the farm accounts, ordering agricultural chemicals, investigating environmental concerns, watching out for farm safety and ensuring the whole farm family were well supported and able to continue their farm activities (participant observation, December 1993 and interview: Sonja, September 1994). Through these activities she was able to weave together her identities and experiences as 'farmer', 'business woman', 'community participant' and 'mother' to highlight issues about emerging marketing challenges, and concerns about chemical usage and farm safety – all of which were concerns held by the movement. This example illustrates the constantly fluid relations between agency and action, where from an experience of multiple positionality, agency is developed with the use of strategic identities and practised through particular actions or mobilisation strategies. These strategies now deserve separate consideration.

## Mobilising strategies

Through the Women in Agriculture movement, farm women have emerged as active players in shaping and changing the politics of Australian agriculture. This is occurring in many 'fields of action' (Liepins 1998a) from the politics of personal identity, through negotiations at the farm level, to industrial participation and power relations at the levels of state, media and industry activity. Action across these scales of politics has occurred as Women in Agriculture participants have employed six main strategies (summarised in Table 8.1). They are listed in order first to identify strategies enhancing recognition and support of women and then to identify strategies explicitly involving engagement with (and challenge to) political

*Table 8.1*    Mobilisation strategies employed by the Women in Agriculture movement, 1993–4

| Strategy | Description | Political implications |
|---|---|---|
| *Strategies enhancing recognition and support of women* | | |
| Becoming 'visible' | Promotion of women in agriculture through public events | Countering women's 'invisibility' in agriculture |
| Developing voice(s) | Establishment of public addresses, events, media releases, and responding to requests for comment | Challenging silence of women's marginal or invisible positions |
| De-reconstructing knowledge | Circulation of women's experiences, knowledge and perspectives via workshops, media releases and industry delegations | Modifying or countering dominant knowledges with additional or alternative truths |
| Developing relations | Formation of networks and supportive connections for exchange of experiences and ideas | Developing sources of support to sustain other actions |
| *Direct political strategies* | | |
| Engaging with dominant political practices | Establishing pragmatic relations and completing due formalities with existing organisations and political systems | Demonstrating acknowledgment of existing political systems and gaining credibility as system participants |
| Reconstructing/creating new political spaces and politics | Exploring alternative forms of decision-making and organisation | Enabling new forms of politics to be established |

arrangements and relations. It should be noted, however, that these strategies do not constitute any form of linear or sequential political activity. Instead, a simultaneous activation of the strategies in different combinations for different situations has enabled the movement to coincidentally promote and enhance women's political participation while also shaping various industrial contexts (such as government institutions and the media). This latter point is discussed in further detail shortly.

## Promotion through public events

A key strategy has involved the promotion of women's involvement and contribution to agriculture. 'Becoming visible' has been a goal in many urban women's movements in the past. It has, however, emerged as a crucial goal for Australian women farmers as they have identified the effects of being previously unseen, forgotten or marginalised as active producers in their industries. Strategies aimed at this goal include the construction of constitutional objectives by various Women in Agriculture groups. For instance, the constitution of AWIA includes two such aims, namely: 'to unite and raise the profile of women in agriculture; ... [and] to gain national and international recognition' (AWIA Minutes, 14 February 1993).

Love (1991: 88) has extended Foucault's (1980) attention to the 'gaze' and 'eye' of power to argue that 'in modern society, individuals are ruled as they are known/seen'. In a similar way, Thiele (1992) has linked visibility and power in her account of how women are often disempowered by strategies that make (or attempt to keep) them unseen and invisible. In the case of Women in Agriculture, groups and leaders explicitly negotiated visibility as a political strategy, believing that to be seen was a key initial goal for more active industry recognition and involvement. Most obvious examples of this strategy included women holding public events, establishing a nationally recognised entity through wide use of the phrase 'Women in Agriculture', seeking government and industry acknowledgement of women's positions in agriculture, and developing broad public awareness of women farmers through the media. For example, the Women on Farms Gatherings have had a constant goal of promoting farm women through the annual conferences. The 1992 Victorian group concluded:

> The Gathering has certainly provided the opportunity for women on farms to raise their profile, to increase their networks and to learn for themselves and their industries. No doubt Tallangatta [the next 1993 Gathering] will once again put women on farms in Victoria on the map!
>
> (Women on Farms Gathering 1992: 7)

Likewise, the 1994 Committee had two objectives related to promoting women as visible contributors in agriculture:

> To acknowledge the ongoing work of rural women;
> To provide increasing recognition of the role of women in agriculture.
>
> (Rural Women's Development Group 1993: n.p.)

## Developing voice(s)

A second strategy mobilised by the Women in Agriculture movement has involved the development of voice(s). Ryan (1991: 206) has argued that power is 'born' in the 'fear of speech'. This view of power as a repressive notion has been subverted within the Women in Agriculture movement as women have explored how the active voicing of their contributions and agendas has been an empowered act – strategically challenging their state of traditional silence and omission with the industry. One participant was to write of this in reference to the Gatherings: 'The Gathering gives us a chance for growth ... we are strengthened in the telling and hearing of our stories' (Field data: Rural Women's Development Group Committee Meeting, 30 March 1994).

Strategies focusing on 'voice' include the widespread incorporation of 'women's stories' in group events and texts, and the external promotion of women's voices and comment in agricultural politics, media and government inquiries.

Strategies based on speaking women *into* agriculture have had a further consequence in enabling women to communicate across their diversity. Love (1991: 96) has argued that 'voice establishes connections across space' enabling the acknowledgement of difference and the possibility of constructing a counterpoint of diverse voices which together constitute a complex but coherent entity. The range of women involved in the Women in Agriculture movement is great, extending from agricultural scientists in government organisations through partners in very large cattle stations, to farmers on medium-sized dairy and mixed farming properties, recent immigrants on small fruit blocks and aboriginal women from collective ownership trusts. It is true that leading activists have tended to emerge from particular class positions within farming; however, the commitment to hear many voices has enabled the movement to span differences and acknowledge collective interests even when women are maintaining diverse commitments and values. For instance, each Gathering based as it is in a different region and town means that farming and other differences are constantly negotiated. In the case of the 1994 Gathering, aboriginal, Scottish, Irish and Dutch women's historic and contemporary experiences were highlighted in presentations of women's stories and farm lives (Rural Women's Development Group 1994b). Likewise, the voicing of collective experiences and interests (for example, issues of farm labour and industry participation) across greatly varying circumstances was evident through the International Conference. This event drew together women from a large array of different farming circumstances (as disparate as subsistence and cash cropping in Kenya, Indonesia and Cambodia, through to large-scale industrial farming in Ireland, Canada and the USA). Yet through processes of debate in 'Action Groups' and workshops, the conference concluded with common generic recommendations regarding women's needs in terms of status such as land tenure and participation in agricultural decision-making processes, which were widely circulated to national governments and the 1995 United Nations Fourth World Conference on Women (Women in Agriculture International Conference Committee 1995a).

## De/reconstructing knowledge

A third strategy mobilised by Women in Agriculture closely relates to women's voices and involves their de/reconstructing knowledge. Participants are supported in circulating and promoting their knowledge through Women in Agriculture newsletters and events. For example, participants' individual 'evaluation' comments from the 1994 Gathering noted that:

• people can help each other by talking about what they have done;
• [a highlight of the Gathering:] passing on knowledge and exchang[ing] views and practices.

(Field data: Post-conference evaluation responses F21, F71)

Likewise, at the International Conference the Governor-General of Australia suggested that '[it is] important in every society that the skills and the experience of women – as well as men – are factored into the perspective of agricultural production', while an activist went further to argue, 'we need to expose our knowledge and own it. We must make our "woman knowledge" important and respected and use [it] to create a new and different experience of agriculture' (Field data: Taped speeches 2 and 3 July 1994).

Foucault (1990) has argued that the construction of knowledge through 'truth-defining' discourses has been instrumental in the articulation and resistance of power. Weedon (1987: 112) has extended this idea to note that maintenance and/or reconstitution of such truths requires the action of agents who are constituted and governed by these discourses. Activist geographies by Mackenzie (1992) and Westwood and Radcliffe (1993) illustrate this process by (respectively) documenting how Canadian farmers and Argentine mothers have appropriated some discourses while modifying the knowledges and subject positions occurring in others. In a similar fashion, the Women in Agriculture movement has recognised and negotiated a range of discourses that shape agriculture. This has included the way in which scientific discourses of agricultural knowledge have been privileged (Kloppenburg 1991) and legal discourses of property and productive contributions have shaped farm ownership and entitlement (Voyce 1994). On both accounts, Women in Agriculture events and texts have contributed to the simultaneous circulation and critical comment of dominant discourses. For instance, common scientific knowledge is increasingly made available to women through workshops and seminars, while additional and alternative farming knowledges have been circulated through the inclusion of workshops and materials on production and value systems that lie outside mainstream farming practices (for example, permaculture, organic and biodynamic farming). Likewise, events within the movement have been important for outlining the legal status and options for farm women, while at the same time activists have been successful in lobbying the federal Attorney-General to direct the Australian Law Reform Commission to investigate the status and experience of women on farms.

## Developing relations

Developing relations has been an unassuming but powerful fourth strategy the Women in Agriculture activists have mobilised for political effect. Acklesberg (1988) has identified how activist networks have a strong function in supporting members in action. Social movements enable participants by mobilising existing affiliations and initiating new ones – both within and external to the activist groups themselves (Gilkes 1988; Goldberg 1991; Crook *et al.* 1992). By developing relations and connections between women, the Women in Agriculture movement has been critical in providing women with a sense of collective support and capacity from which to work outwards and tackle the agricultural and social agendas they have chosen. For key activists in the movement, committee and planning meetings have been crucial events for drawing women together from quite distant and different physical and social contexts. One activist explained, 'The meetings are incredibly important for support. It's just simply knowing that there are other women with like minds and like aims and goals' (Kay, September 1994). Similarly, for general participants, the power of conferences, group meetings and newsletters has been vital for women sparsely scattered throughout Australia who came to feel a sense of connection with others of like interests. One participant described the impact of the Gatherings:

> The only equivalent I would say is the sort of openness you get ... if you're travelling in a foreign country and you haven't met anyone who speaks your language for a few days, and you meet someone who's from your country and suddenly you have this rapport.
>
> (Pam, January 1994)

Another Women in Agriculture participant said: 'It is a great boost to know there are lots of others similar to me [i.e. a self-defined farmer], so many here-abouts [in the local area] see themselves purely as housekeepers' (Field data: Post-conference evaluation response E90).

## Engaging with dominant political practices

The fifth and sixth strategies frequently mobilised by the Women in Agriculture movement involve actions which directly engage or challenge existing political arrangements (see Table 8.1). In the first case, participants of the movement individually and collectively engage with dominant practices of politics. This involves a degree of 'opting in' to mainstream political systems which women wish to change. Gilkes (1988: 68) argues that activism necessitates building strategic or 'pragmatic affiliations' whereby political relations and institutional processes are engaged and partially supported by activists in an effort to change power relations and agendas 'from within'. In this fashion, participants and activists from Women in Agriculture have individually and collectively established ongoing relations with mainstream farmer organisations, industry boards and government departments. These relations are built on a degree of recognition and acceptance of political and procedural systems that the movement accords such bodies. Consequently, the organisations then find it less problematic to receive the submissions and lobbying activities mounted by the movement. Examples of these strategies include the good relations groups established with the staff of the federal Department of Primary Industries and Energy, and the Victorian Department of Agriculture. Results of such strategies were crucial to both the financial resourcing and political recognition of the movement during the 1990s. At state, national and international levels, the movement's events have received major funding from government and industry sources. Moreover, the cumulative lobbying by the movement led to the establishment in 1995 of the federal Rural Women's Unit within the Department of Primary Industries and Energy. Likewise, consultation with Women in Agriculture groups was increasingly sought by farmer organisations and government institutions throughout the mid-1990s and even subsequent structural changes were undertaken by some organisations to increase the participation of women.[5]

## New political spaces and politics

The second case of direct political strategies mobilised by Women in Agriculture involves the movement's reconstruction and/or creation of new practices and power relations. In part, this involves exploration of the political and social space of the movement as one of 'new', different and unbounded opportunities where meanings, power relations and political practices can be pushed beyond current forms. In this regard, new social movements have been identified for their ability to maintain scepticism and alternative practices to bureaucratic institutions and conventional forms of debate, decision-making and politics (Crook *et al.* 1992; Rosenau 1992; Melucci 1996). Women in Agriculture demonstrates such opportunities through its creation of new learning environments for women, alternative agricultural texts and different forms of information sharing and political action. A description of the supportive multi-layered environment of the Gathering illustrates this situation: 'It's fairly frank, very open. There's an understanding of what you can talk about. It can be from the personal to the technical to the political ... all of these things interact together' (Catherine, January 1994).

The political space of the Gathering is therefore a stark contrast to the hierarchical, procedural and excluding experience many women noted of their local Farmers' Federation meetings.

> It [Women in Agriculture] is very different from the Feds where there's this boys' sort of network and there's established routine, 'You support me in this motion and I'll support you in yours even if I don't feel that comfortable with it.' That's simplifying it all but that is the way those groups operate and there's ... a set of rules that women don't know about and aren't comfortable with.
>
> (Kay, June 1994)

The combined assemblage of strategies mobilised by Women in Agriculture has influenced participants' work for individual and collective recognition of their positions and contributions in agriculture. Strategies concerning being seen, gaining voice(s), de/reconstructing knowledge and developing relations have all aimed at enhancing the acknowledgement of women in farming. These first four strategies have enabled women to work for such recognition at individual levels on their own farms and in their own communities. Women have used them, both individually and simultaneously. For instance, the committees of the Gathering, AWIA and the International Conference have all taken opportunity through media and government meetings to simultaneously 'be seen', raise their voices and express their own knowledges of agriculture through media releases or presentations to government. While much of the focus of these first four strategies is aimed at women's positions in agriculture, the two final strategies combined with these earlier ones provide women with political relations and practices to shape and challenge the social, economic and political systems that construct and govern agriculture. It is the spaces and effects of activism at these two different levels of politics (the personal and the industry-wide) that form the final section of this chapter.

## Spaces and effects of Women in Agriculture

*Kate is a young woman farming a large grain property with her husband. She has several young children to care for while also being actively involved in all aspects of the farm business. Kate had experienced a sense of isolation in her local community because 'so many women here keep to their family and home and community activities, or they're interested in their golf or tennis'. After attending the International Conference, Kate returned home with a strong affirmation of her own preferred identity as 'a farmer'. She then went on to individually and collectively organise a number of initiatives:*

> It was like an energising experience. I came back and organised a few things differently here on the farm. But I also kept in contact with women I met through Women in Agriculture. I talked to our local paper and then worked with the others to get our state network off the ground. I also went along with [Grace] to the [branch farmer organisation] and started participating there. They need to change the way they do things, and we can make a start there together.

The Women in Agriculture movement has 'made a space' for women in the industry. A detailed reading of all the sites, scales and spheres of Women in Agriculture politics is beyond the scope of the current chapter (see Fincher and Panelli 2001; Liepins 1996a, 1998a). However, the effect of this agency has been profoundly geographic in both a material and a social sense. As Kate's experience illustrates, results have ranged from changes in women's individual positions, identities and relations through to the physical organisation of their farms and the social and material changes they have achieved within farming, media and government organisations. Groups and individuals within the movement have selected and promoted key identities as farmers, business partners and mothers (Liepins 1998c) and from these positions they have employed the strategies discussed above to gain recognition and participation at both personal and industry levels.

At the personal level, Women in Agriculture participants have reported a variety of changes in their individual relations and farm arrangements as a result of the movement. Individually, many women recorded a sense of affirmation for their contributions in farming and their opportunity to voluntarily select and project an identity as farmers. Elaine was to describe and echo the experience of many women participating in Women in Agriculture when she spoke at the 1992 Gathering saying, 'I went to Sea Lake [site of the 1991 Gathering] a farmer's wife and came back home a farmer' (Field data: Women on Farms Gathering 1992 video tape footage taken at Numurkah). This statement encapsulated the process of personal 'reconstruction' that many participants have recorded in evaluations and newsletters circulated through the movement. In a very real sense, the movement has been important to women's abilities to reposition themselves socially and economically within their farm environment.

While some of these effects rest with the personal choices women make about how they represent themselves, other equally significant results followed as women chose to operate differently back on their farms and in their local communities. Participants at the 1994 Gathering were asked about the impact of the event for them as farm women. A crop and wool farmer noted, 'I understand more about the working of the farm', while a beef farmer highlighted the Gathering as '[giving] me the confidence in doing things myself', and a dairy farmer reported she 'will put into practice more' (Field data: Gathering evaluation responses F13, E08, F80). The most comprehensive documentation of these types of effects occurred through the two-part evaluation activity, conducted both at the close of the International Conference and three months later. Farm level changes and increased farm and community involvement were very commonly reported:

> The conference has given me confidence to play an even more active role in agriculture, particularly on my own farm.

> The confidence (personal) I drew from the conference enabled me to become more *vocally* [own emphasis] involved on the farm.

> I have asserted my ability to do activities [on a grain farm] usually done by the men. If brawn can't do it, I have to use my brain!

> [Three months on] I try to accept more tasks which I used to always leave to my husband – e.g. learning new skills and fixing things.

> After the conference I reorganised the book keeping, stock records, [and] contacted Primary Industries concerning plant tissue testing and blood testing stock.

[Three months on] I have increased my confidence, I've taken a role in a vital community issue, done things which I would have shunned before.
(Field data: Post-conference evaluation responses E81, E41, E73, F60, E15, F89)

Results such as these illustrate that across the movement the effect of ongoing group events and networking newsletters has a significant role in changing the positions and opportunities for women both in their personal farming arrangements and broader commitments. In a material sense through work and financial arrangements, and in a social sense through family relations, communication and decision-making processes, Women in Agriculture is a movement that is reshaping the spaces and ways in which women can participate in agriculture.

Beyond the immediate scale of personal farm units and local community/farming issues, Women in Agriculture has also created a political space and a series of results for women in a variety of 'industry' arenas. These include farmer organisations, rural/agricultural media, commodity boards, and primary industry departments at different levels of the government. In particular, the movement has had three main effects. First, activists in the movement have mobilised their 'farmer' and 'business partner' identities to secure recognition for women's contributions to farming. This acknowledgement has been sought and gained in funding or rhetorical affirmation of the movement's activities from commodity marketing and regulation boards (for example, Australian Wheat Board and the Victorian Dairy Industry), farmer organisations (for example, Victorian Farmers Federation and National Farmers Federation) and parliamentarians and government bodies (including the federal Department of Primary Industries and Energy, Office of Women's Affairs and the Rural Education Access Programme, the Victorian Department of Agriculture, and Occupational Health and Safety Authority). Philanthropic and commercial organisations have also played a part in acknowledging, sponsoring and publicising the movement (Liepins 1996a). By gaining the support of these types of organisations, the movement has secured financial resources and public statements of support for its goal to have women recognised as important and multi-skilled contributors to agriculture.

A second result produced by the Women in Agriculture movement across a number of industry arenas involves the gains in access and influence with key organisations. These achievements relate directly to the strategies activists have mobilised in developing women's voice/s and knowledge in agriculture. Participants in the movement have reported through newsletters and evaluation activities that Women in Agriculture events have encouraged them to more actively participate in farming organisations, industry bodies and community fora:

[The Conference] got me more politically aware that I have a voice and some very good ideas.
(Women in Agriculture International Conference Committee 1995b: 49)

The Gathering has given me greater confidence in my ability to argue a point in group discussions.

> (Field data: Gathering evaluation response E17)

[The Conference gave me] greater credibility, enabling me to be more effective in lobbying for change [as] people listen to me more than before.

> (Field data: Post-conference evaluation response G13)

I am involving myself in the political arena ... being proud to voice my opinion because of my own experience, and encourag[ing] others who work the land that their voice is of value too.

> (Women in Agriculture International Conference Committee 1995a: 269)

Since the Conference [I have been] agreeing to make speeches which gave me a shock as I would never have done it before the Conference.

> (Women in Agriculture International Conference Committee 1995b: 44)

While these results indicate the range of individual actions participants have taken, collective accessing of industry fora and knowledge production has been equally effective. For instance, the committee organising the 1994 Gathering corresponded with the Victorian Farmers Federation criticising that organisation's sexist public statements and advice about farm transfer. This was seriously received by the Federation and visits were exchanged, and adjustments to policy and texts ensued.

Similar recognition of the movement as a source of women's voices and knowledge in agriculture was noted in government organisations seeking Women in Agriculture representation on consultation panels and policy initiatives.[6] Likewise, the movement's opinions and perspectives have been widely recorded and circulated through a range of media. This included rural and metropolitan electronic and print media covering the groups' events and agendas, seeking comment on feature issues, and initiating projects to increase farm women's voices in the media. For instance, the largest rural newspapers in Australia, *The Weekly Times* and *Stock and Land*, established regular farm women's columns which were co-administered by activists in the movement. Likewise, the Australian Broadcasting Commission funded an ongoing, nation-wide Rural Woman of the Year Award commencing with the 1994 inaugural gala night as part of the International Conference Banquet. Groups within the movement also established assertive media campaigns which were well received and supported by different media organisations (Liepins 1996b).

The third widely achieved industry effect achieved by the Women in Agriculture movement involves the ongoing development of networks and political affiliations. This has resulted from the movement's pragmatic strategy of developing relations, discussed earlier. In the case of farming organisations, the National Farmers Federation (NFF) received and encouraged correspondence with AWIA following the success of the International Conference. Relations warmed in the following year to the point where a joint media release was issued stating:

> NFF and AWIA are investigating ways to jointly promote the positive role
> that women play in agriculture and rural communities ... As a result of
> today's discussions, NFF and AWIA have agreed to hold further talks on
> ways to promote the role of women in Australian agriculture and farm
> organisations.
>
> (NFF–AWIA 1995: n.p.)

At a government level, both movement- and government-initiated relations
have been established and maintained at state and federal levels through the
various portfolios of women's affairs and primary industries. For instance, the
Foundation for Australian Agricultural Women (a body emerging from the
International Conference) was closely consulted by the federal Australian Bureau
of Agriculture and Resource Economics (ABARE) (ABARE 1995) when a
national statistical study of *Women on Farms* was commissioned by the government. Likewise, submissions have been encouraged by Senate inquiries and the
Australian Law Reform Commission on issues relating to the experience and
status of women on farms (Liepins 1996a).

Cumulatively, then, the movement has formed an alternative space from
which to work outwards from participants' personal positions. The movement
encourages an overlapping of its own political space with the personal and
industry spheres to which the women have wished to contribute. Thus the movement has provided a political space for women to work from their previously
marginal positions to create agricultural connections and opportunities – on
their farms, through media, industry boards, farmer organisations and government bodies.

## Conclusion

The Women in Agriculture movement is a story of individual women like Gwen,
Brenda, Maureen, Pam, Sonja, Kay, Elaine, Catherine and Kate. It is also a
story of group strategies, mobilising across great diversity to effect changes at a
number of scales. This is not a story of total harmony or unity. Indeed, contradictions and tensions have arisen because women have come from contrasting
backgrounds and with varied resources. These circumstances have meant that
participants and leaders have often held quite different perspectives on issues: for
example, their varied positions on environmental issues associated with agriculture (see Liepins 1998c).[7] Nevertheless, across women's multiple positions in
farming, they have jointly experienced prejudice or constraints concerning their
recognition as farmers and their access to industry arenas. These experiences
have stimulated their collective agency in a bid for greater visibility and participation in their industries. Interestingly, the results of the Women in Agriculture
movement also reflect Cope's (1997: 95) key concepts of a gender-sensitive political geography, namely politics surrounding 'participation, power, and policy'.
The scope of Women in Agriculture politics has affected women's participation
in their family, farm and industry settings. So, too, it has supported women's

exploration and negotiation of power relations in a range of personal, industry and public spaces. Finally, the movement has included targeted actions aimed at policy arenas of both state and federal governments as well as farmer organisations. Together, these achievements have succeeded in writing women into farming and agricultural industries, through a politics that explicitly articulates strategic identities, voices and relations.

## Notes

1   Ruth Panelli has formerly published under the name Ruth Liepins.
2   I have learnt much from, and been greatly inspired by, the range of women active in Australian farm politics. Both for methodological and political reasons I have wished to write some of these women *into* my chapter. I do this because it symbolises the continual interrelations I have had with the movement based on a mutual learning and constructing of each other's experiences and understandings about Australian agricultural politics and the representation of that politics. On the one hand, activists were (among other things) farmers and agricultural workers who often represented me as, and at times looked to me to be, a researcher and articulate writer. On the other hand, I worked with the movement as a feminist and a doctoral student who often represented the women as, and at times expected them to be, farmers and activists. Throughout our interaction, our joint activism enabled us to learn much about both Australian agriculture and the importance of choosing thoughtful constructions of our politics. I now include 'cameos' of a selection of women in this chapter, noting explicitly that such narratives are of my own creation, but also aiming to acknowledge their interest in circulating women's stories and paying respect to their own accounts of their agricultural situations and political actions. Throughout the text, real names are used where women have had their stories published. Pseudonyms are employed to maintain anonymity where women have recounted their experiences more privately.
3   A form of action research was established with three groups working at different scales in the movement. Originally, a partnership research relationship was sought with the movement, influenced by my reading about participatory action research methodologies. However, the majority of women's energies were expended on the political work of the movement and their ongoing economic and social responsibilities in farming, thus the engagement with a participatory research process was usually transitory in terms of the priorities they chose and met.
4   The movement's roots can be traced to two early 'one-off' conferences staged for rural and farm women in Victoria during the late 1970s and early 1980s. Momentum then built during the 1980s as women's contributions on farms and in rural communities were featured during some severe drought conditions affecting the state. By the late 1980s, state government-funded farm skills courses were being run for women and a rural women's network was established (1986). The Gathering, AWIA and International Conference all originated from within Victoria. These types of activities and networks were to spread throughout other states during the 1990s, including the establishment of regular statewide Gatherings, attendance at the 1994 International Conference by women across Australia, and involvement in the national group, AWIA.
5   For example, the Victorian Farmers Federation voted in 1995 to change its constitution to enable two farm votes per farm membership. The executive officer explained this change as heralding 'a new era in encouraging greater participation of women and young people' (Hunt 1995:1).
6   In 1994–5, these developments included seeking 'Women in Agriculture' perspectives on federal landcare policy, the Victorian Drought Information and Support Project, Australian Bureau of Agricultural and Resource Economics projects and the Office of

the Status Women's Australian reports to the United Nations Fourth World Conference on Women.

7   A further example of tensions includes the different experiences and opportunities women had in participating in both agriculture and the Women in Agriculture movement, dependent on whether they were married, in a *de facto* relationship, widowed or single. These differences are beyond the scope of the current chapter but deserve further consideration.

# References

Acklesberg, M.A. (1988) 'Communities, resistance and women's activism: some implications for a democratic polity', in A. Bookman and S. Morgen (eds) *Women and the Politics of Empowerment*, Philadelphia: Temple University Press.

Australian Bureau of Agriculture and Resource Economics (ABARE) (1995) *Women on Farms: A Survey of Women on Australian Broadacre and Dairy Family Farms 1993–1994*, Canberra: ABARE.

Australian Women in Agriculture (AWIA) (1993) Minutes, 14 February.

Cope, M. (1997) 'Participation, power and policy: developing a gender-sensitive political geography', *Journal of Geography* 96, 2: 91–7.

Crook, S., Pakulski, J. and Waters, M. (1992) *Postmodernization: Change in Advanced Society*, London: Sage.

Davies, B. (1991) 'The concept of agency: a feminist poststructuralist analysis', *Social Analysis* 30: 42–53.

Fincher, R. and Panelli, R. (2001) 'Making space: women's urban and rural activism and the Australian state', *Gender, Place and Culture* 8, 2: 129–48

Foucault, M. (1980) 'The eye of power', in C. Gordon (ed.) *Power/knowledge: Selected Interviews and Other Writings 1972–1977 by Michel Foucault*, New York: Pantheon Books.

——(1990) *The History of Sexuality: Volume 1*, London: Penguin.

Gilkes, C.T. (1988) 'Building in many places: multiple commitments and ideologies in black women's community work', in A. Bookman and S. Morgen (eds) *Women and the Politics of Empowerment*, Philadelphia: Temple University Press.

Goldberg, R. (1991) *Grass-Roots Resistance: Social Movements in Twentieth Century America*, Belmont: Wadsworth Publishing.

Hunt, P. (1995) 'Women given the vote', *The Weekly Times*, Melbourne, 5 July.

Kloppenburg, J. (1991) 'Social theory and the de/reconstruction of agricultural science: local knowledge for an alternative agriculture', *Rural Sociology* 56, 4: 519–48.

Kyne, T. (1994) 'Women in fight for land rights', *Herald Sun*, Melbourne, 24 March.

Liepins, R. (1996a) ' "Women in agriculture": a geography of Australian agricultural activism', unpublished PhD thesis, University of Melbourne, Australia.

——(1996b) 'Reading agricultural power: media as sites and processes in the construction of meaning', *New Zealand Geographer* 52, 2: 3–10.

(1998a) 'Fields of action: Australian women's agricultural activism in the 1990s', *Rural Sociology* 63, 1: 128–56.

——(1998b) 'The gendering of farming and agricultural politics: a matter of discourse and power', *Australian Geographer* 29, 3: 371–88.

——(1998c) ' "Women of broad vision": nature and gender in the environmental activism of Australia's "Women in Agriculture" movement', *Environment and Planning A* 30: 1179–96.

Love, N.S. (1991) 'Politics and voice(s): an empowerment/knowledge regime', *Differences* 3, 1: 85–103.

Mackenzie, F. (1992) '"The worse it got the more we laughed": a discourse of resistance among farmers of Eastern Ontario', *Environment and Planning D: Society and Space* 10, 6: 691–713.

Melucci, A. (1996) *Challenging Codes: Collective Action in the Information Age*, Cambridge: Cambridge University Press.

Mohanty, C.T. (1995) 'Feminist encounters: locating the politics of experience', in L. Nicholson and S. Seiman (eds) *Social Postmodernism: Beyond Identity Politics*, Cambridge: Cambridge University Press.

Mouffe, C. (1995) 'Feminism, citizenship and radical democratic politics', in L. Nicholson and S. Seiman (eds) *Social Postmodernism: Beyond Identity Politics*, Cambridge: Cambridge University Press.

National Farmers Federation (NFF)–AWIA (1995) *NFF–AWIA Joint Media Release*, Canberra, 12 July.

Rosenau, P.M. (1992) *Post-Modernism and the Social Sciences: Insights, Inroads and Intrusions*, Princeton: Princeton University Press.

Rural Women's Development Group (1993) *Women on Farms Gathering: Sponsorship Proposal*, Glenormiston: Rural Women's Development Group.

——(1994a) *Women on Farms Gathering: Programme*, Glenormiston: Rural Women's Development Group.

——(1994b) *Women on Farms Gathering: Proceedings*, Glenormiston: Rural Women's Development Group.

Ryan, M. (1991) *Politics and Culture: Working Hypotheses for a Post-Revolutionary Society*, Basingstoke: Macmillan.

Thiele, B. (1992) 'Vanishing acts in social and political thought: tricks of the trade', in L. McDowell and R. Pringle (eds) *Defining Women: Social Institutions and Gender Divisions*, Cambridge: Polity Press and Open University.

Voyce, M.B. (1994) 'Testamentary freedom, patriarchy and inheritance of the family farm in Australia', *Sociologia Ruralis* 34, 1: 71–83.

Weedon, C. (1987) *Feminist Practice and Post-structuralist Theory*, Oxford: Blackwell.

Westwood, S. and Radcliffe, S.A. (1993) 'Gender, racism and the politics of identities in Latin America', in S.A. Radcliffe and S. Westwood (eds) *"Viva": Women and Popular Protest in Latin America*, London: Routledge.

Women in Agriculture International Conference Committee (1995a) *1994 International Women in Agriculture Conference – Farming for our Future*, Sale: Women in Agriculture International Conference Committee.

——(1995b) *Agents for Change: Evaluation of the First International Women in Agriculture Conference, Melbourne, 1994*, 'Sale: Women in Agriculture International Conference Committee.

Women on Farms Gathering (1990) 'Women on Farms Gathering proceedings', unpublished proceedings collated by Women on Farms Gathering.

——(1992) 'Women on Farms Gathering proceedings', unpublished proceedings collated by Women on Farms Gathering.

# 9 The complexities of women's agency in Fiji

*Jacqueline Leckie*

This chapter examines the complexities of women's agencies and activism in Fiji. Although the most isolated island communities are not immune to the impact of globalisation, women negotiate this within cultures where tradition, religion and the legacy of colonialism remain strong. However, historical, cultural and spatial specificities complicate an understanding of women's agencies and alert us to the dilemma of treating the category 'women' either as unified or as infinitely fragmented. Theories that mediate between these extremes, such as seriality (Young 1995) and multiple identities (for example, Brunt 1989; Moghadam 1994; Chhachhi and Pittin 1996) are considered against the context of gender agencies in Fiji. This is discussed in four sections. The first section examines from both theoretical and empirical perspectives how agency is gendered, *viz* is women's agency different from men's?

The second section then focuses on how agency transposes into active negotiation. What are the linkages and disengagement between the state, work and unions, church, village, non-governmental organisations (NGOs) and more informal community and household networks – the messy realities of everyday activism? How does covert relate to overt agency? Third, the chapter explores the specificity of place, culture and representations in Fiji and the conundrum of essentialising women and representation. The fourth section reassesses whether the theories of seriality or multiple linkages offer useful explanations of the renewed assertiveness of women in Fiji and/or simply expose the differing impact of constraints on women today (Leckie 1997a). Before these issues are examined, the next section contextualises the discussion in Fiji.

## Context

Fiji emerged as a sovereign state after substantial transformations under British colonial rule between 1874 and 1970 (Lal 1992). More recently, since the late 1980s, dramatic political and economic restructuring (Robertson 1998) from internal and global forces have set Fiji apart from its island neighbours.

Fiji's colonial economy was based upon the export of sugar which led to the immigration of approximately 60,965 indentured workers (*Girmitiyas*) from India between 1886 and 1916. Approximately 30 per cent were female migrants

(Shameem 1990b: 5) who constituted both cheap plantation and unpaid domestic labour, especially after the shift from large plantations to small family farms. The growth of the latter was fostered by the colonial sugar refining company (CSR) after indenture was abolished in 1916. Large plantations became unprofitable partly due to a shortage of labour. CSR maintained its advantage by leasing land to Indo-Fijian farmers, controlling many of the conditions of cane production, prices and the provision of credit facilities (Sutherland 1992: 34–8). Indo-Fijians also leased land from indigenous Fijians. The colonial state attempted to segregate the two main ethnic groups so that Fijians were predominately confined to their villages. As discussed below, this had significant gendered implications as colonial policies, Christianity and 'chiefly hierarchies'[1] constricted the options for most indigenous women.

Fiji has developed a more industrialised and urbanised economy compared to many Pacific states, although the subsistence sector is still pervasive on the hundreds of scattered islands in Fiji. A larger and multi-ethnic proletariat developed (Leckie 1990). Within the present population of around 800,000, the two main ethnic groups of indigenous Fijians and Indo-Fijians are fractured by economic inequalities. Europeans and transnational companies command considerable wealth. Class, wealth, status and power are not necessarily conterminous. For example, an indigenous Fijian community might have extensive land rights but be relatively poor although their land is leased to Indo-Fijian families. The incomes of Indo-Fijians vary and have been shaped more by educational and occupational opportunities. These factors are also increasingly vital to indigenous Fijians but traditional status, especially between chiefs and commoners, and different geographical regions are also determinant. Age is another variable that translates into status, power and material resources.

Gender cuts through the aforementioned currents which converge in modern Fiji. The problem is that either ethnicity or class has been prioritised in political analysis (Sutherland 1992), particularly in accounting for two military coups of an elected social democratic government in 1987. A popular but misguided stereotype was that the coups were caused by fears of Indian political and economic domination. This focus obscured economic inequalities (UNDP 1997) and other political undercurrents, too complex to delve into here (for details, see Robertson 1998).[2]

Closer examination also reveals a marked gendered dimension to economic and political changes in Fiji. Post-coup recovery strategies promoted export-orientated industrialisation that depended upon cheap female labour (Slatter 1987; Emberson-Bain 1992: 151–3; Harrington 1999). Women have been susceptible to poverty, both through their concentration as low income earners and also because of an increasing number of poor female-headed households (UNDP/Government of Fiji, 1997: 76–7). The introduction of measures such as indirect taxes and market rentals for state-assisted housing, along with currency devaluations in 1987 and 1998, has contributed to women's economic insecurity. Although the coups supposedly represented a political revolution for indigenous Fijians, this did not translate across gender lines. A major survey (Booth 1994) reiterated the lack of women from all ethnic groups at higher levels in political and economic organisations in Fiji. Violence against women has also constituted another serious problem (Emberson-Bain 1992: 153–5).

Predictions were cast that gender politics and women's agency would become swamped under the prominence of ethnic politics during the post-coup period:

> the women's movement suffered a major setback, particularly in its attempts to mobilise and unite women on the basis of gender irrespective of race … The coups and the attempts to incite ethnic polarisation have important implications and ramifications for women, particularly in terms of the mobilisation and actions they mount as women. Essentially, despite their common gender, ethnic Fijian and ethnic Indian women are divided, or at least have the potential to be divided, on the basis of ethnicity.
>
> (Lateef 1990a: 125)

This forecast proved excessively gloomy as women's activism was reasserted much more strongly after than before the coups. What strategies did women pursue to negotiate the tumultuous post-coup period? How has the heterogeneity of other identities, such as class/ideology/status/tradition/age/sexuality/ethnicity/religion, presented strengths and weaknesses for alliances between women? (See Yeatman 1993; Yuval-Davis 1994; Chhachhi and Pittin 1996.) In Fiji, ethnicity, nationalism and tradition are especially powerful identities intersecting with gender.

## Women's agency

The theorisation of agency has been problematic partly because this straddles structuralist and more subjective post-structural identities. Several writers have theorised agency in relation to time and space; much of this equates to Homi Bhabha's (1994) discursive 'third space' where political activism accepts difference but does not necessitate essentialist (and hegemonic) structures or identities.

Although feminist politics has not depended upon a unitary concept of woman (for example, Mouffe 1993), there are still problems with the theorisation of gender and agency. Chhachhi and Pittin (1996: 93) noted this gap in relation to the articulation of women workers' consciousness. The issue of an agency specific to women leads into a quagmire of essentialisms about global sisterhood, and the 'Third World' woman (for example, Mohanty 1988; and specifically of 'the Maori woman', see Matahaere-Atariki 1998), and can gravitate towards a universal trajectory of resistance for female subalterns to embark upon.

Iris Marion Young (1995) has theorised the dilemma of gender and agency by resurrecting Jean-Paul Sartre's concept of seriality as a description of unorganised class existence. This distinguishes between social groups and social series. Young (1995: 199) applies the term 'group' to 'the self-consciously, mutually acknowledging collective with a self-conscious purpose' but not all structured social action occurs in groups. This is where the concept of a social series is useful, referring to a collective which does not necessarily have shared attributes or cohesive identity but does have a loose unity, described as a shared passive relationship to a material milieu. The latter shapes individual agency but is not deterministic. Women are positioned in a series 'through the material organisation of social relations as enabled and constrained by the structural relations of enforced heterosexuality and the sexual division of labour' (Young 1995: 208). Young (1995: 207) argues that:

> the concept of seriality provides a useful way of thinking about the relationship of race, class, gender, and other collective structures, to the individual person. If these are each forms of seriality, then they do not necessarily define the identity of individuals and do not necessarily name attributes they share with others. They are material structures arising from people's historically congealed institutionalised actions and expectations that position and limit individuals in determinate ways that they must deal with … individuals can relate to these social positionings in different ways; the same person may relate to them in different ways in different social contexts or at different times in their lives.

The concept of seriality is pertinent to a discussion of women's agency in Fiji where both individuals and groups must constantly negotiate change for women with their identity as members of an ethnic group, local community, union or religious group. On the one hand, these serial memberships may not be pertinent to a woman's sense of identity; on the other, 'her family, neighbourhood, and church network makes the serial facts of race, for example, important for her identity and development of a group solidarity' (Young 1995: 207).

Chandra Mohanty (cited in Nicholson and Seidman 1995: 12) prefers to emphasise the outcomes in women's political agency as 'a notion of political struggle where alliances are made around explicit goals rather than presumed on the basis of imputed commonalities. It represents "sisterhood" as that which needs to be achieved rather than simply being assumed.' This focus on goals that improve women's conditions has been pivotal in the success of more recently established women's groups in Fiji. These groups tread the controversial path of recognising difference – including traditional conservative identities – among women, but also of fostering a common feminist consciousness. The following section will tackle three such patterns that have arisen in the shaping of women's agency in Fiji: (a) issues of sile; (b) the salience of the personal and politics of the body to women's agency; and (c) the multiple axes and 'messy realities' of women's agency and activism (Chhachhi and Pittin 1996). These patterns are indicative of the structural milieu within which women find themselves sharing commonalities, as in a 'series' rather than a unified activist group (Young 1995: 210–11).

### Silence and agency

Women's omission from public and, more precisely, political histories has been one rationalisation for their continued subordination and absence from contemporary politics. Partly to break this silence, some activists and academics are reconstructing the hidden record of women's agency, as individuals[3] and collectively, through women's village, religious and other community organisations (for example, Schoeffel 1979; Kikau 1986; Thomas 1986). Representations of Pacific Island women as passive subordinates, mutely confined to domestic roles, have also been challenged (Linnekin 1990; Ralston 1992: 168). Vicky Lukere (1997: 66) found that Charles Wilkes' (1845) observations in 1840 of the district of Macuata in Fiji 'abound with references to women who wielded great power in their own right' but that Christianity, codified law and colonial rule eroded the power of women chiefs. However, women may have been able to express a 'kind of subalternity' within ritualised contexts and the new religion, Christianity. 'The desire of women to follow the *lotu* [church] often involved rebellion against their husbands' (Lukere 1997: 80).

Arguably, just as the void in representations of women's agency may have reinforced the marginalisation of expressions of activism, so reclaiming knowledge of Pacific Island women's status, power, resources and agency can be empowering for present and future agency. Yet there are many contesting readings of how women's history constitutes agency. The very sites of past female agency, such as the Christian Church, could equally restrain assertiveness.

### Agency and the personal: the body

The connection between the personal and the political has long been a rallying cry in western feminism and saw a postmodernist revival with discourse on the politics of the body, albeit the problems of biological essentialism and possibilities for women's agency (for example, Nicholson 1995). Yet if we are to consider how women have negotiated the boundaries between the individual and the collective, the public and the private, and the political and the personal, then it seems vital to address the connections between agency and the body. Young (1995: 204) emphasises the milieu of enforced heterosexuality as a fundamental structure of women's seriality.

The connection between agency and the body is poignant for most women and permeates through to reproductive choices (Jolly 1998a: 1). Lukere (cited in Jolly 1998b: 193–4, 207–8) has documented the draconian measures taken by the colonial state and male chiefs to constrain the choices, activities and movements of women in Fiji. If a woman was absent from her *koro* (village) for more than sixty days in 1912, she could be fined (cited in Jolly 1998b: 195). Regulations did not just limit women's mobility or work but extended to reproduction, as through various hygiene missions. Women did resist such controls, including prohibitions on abortion. *The Report of the Commission Appointed to Inquire into the Decrease of the Native Population 1893–1896* refers to a 'freemasonry among women which conceals the practice [of abortion] not only from the police but from their husbands and fathers' (cited by Lukere, in Jolly 1998b: 207).

Margaret Jolly, among others, has done much to bring the politics of the maternal subject to academic feminist discourse of the Pacific. This project has centred around the role of the mother. In contrast to many western feminists, 'Asian and Pacific feminists have more readily embraced the maternal subject position, although often to distinguish themselves from what are perceived as anti-family tendencies in Western feminism or as part of anti-colonial or nationalist movements' (Jolly 1998a: 2).

Pacific Island women's agency, activism and identity clearly embrace the maternal and family. Pacific Island women turned discourses of being bad mothers around to being very good ones, even better than white women (Jolly 1998b: 202). This connection between the maternal and women's agency applies to both conservative and radical women's organisations in Fiji (see Table 9.1).

The Soqosoqo Vakamarama, founded in 1924 as a Methodist organisation to foster Fijian women's crafts and introduce modern western home-making skills, became the main organisation for Fijian women (Kikau 1986). Missionaries and the colonial state encouraged such 'native agency' to uplift or 'improve' indigenous women (Lukere 1997: 149–54; Haggis 1998; Jolly 1998a: 11). Today the Soqosoqo Vakamarama is government-funded and all Fijian women automatically become members when they turn 16 years old (Geraghty 1997: 14). It has taken a significant role in post-coup Fiji by emphasising the maintenance of Fijian cultural traditions and crafts, and women's roles as wives and mothers. The Soqosoqo Vakamarama is identified with traditional chiefly leadership. Fijian nationalism, like other conservative anti-systemic movements, invokes traditional household and patriarchal forms to proclaim unique space in the global arena (Smith 1994: 38).

Radical women's groups in Fiji, principally the Fiji Women's Crisis Centre (FWCC) and the Fiji Women's Rights Movement (FWRM), also connect women's agency to the politics of the body. They stress the protection of women and children but within a feminist agenda that is empowering and takes a proactive anti-violence stand. Recently, conservative and radical women's groups have met on common ground over concern about family and women's welfare. For example, the Fiji Methodist Women's Fellowship conducted a workshop with the FWRM on legal literacy. Methodist women leaders were challenged[4] to consider women's rights as part of liberation theology and to 'break the "culture" of silence over the violation of their bodies so that the truth can set us free' (*Balance* July–August 1997: 8). During 1998 the Soqosoqo Vakamarama launched a book in the Fijian language, *Na i Vola Dusidusi Me Vukea Na Kena Karoni Na Matavuvale* ('How to Protect the Family'). Subjects covered included motherhood, pregnancy, contraception, hygiene, parenthood, sexually transmitted diseases (STDs), human immuno-deficiency virus (HIV), acquired immune deficiency syndrome (AIDS) and abortion.

*Table 9.1*  Selected women's groups in Fiji

| Name | Acronym | Year founded | Ethnicity | Focus |
|---|---|---|---|---|
| Ministry of Women and Culture | – | 1987 | Mixed | Government |
| Fiji National Council of Women | FNCW | 1968 | Mixed (Fijian dominant) | Umbrella organisation of women's groups |
| Young Women's Christian Association of Fiji | YWCA | 1962 | Mixed | Health, environment, rural development |
| Women in Politics in the Pacific Centre | WIPPAC | 1995 | Mixed | Political training |
| Women in Politics | WIP | 1998 | Mixed | Sub-project of WIPPAC |
| Women's Coalition for Women's Citizenship Rights | WCFCR | 1996 | Mixed | Political, human rights |
| Soqosoqo Vakamarama | – | 1924 | Fijian | Crafts, homemaking, development work |
| Methodist Women's Fellowship | – | 1948 | Mixed (Fijian dominant) | Religious, crafts, self-help |
| Lautoka Sammlit Women's Club | – | 1979 | Indo-Fijian | Co-ordinate women's clubs, development projects, crafts |
| Ra Nari Kaliyan Parishad | – | | Indo-Fijian | Rural women's clubs |
| Democratic Women's Forum | – | 1990 | Indo-Fijian | Pro-democracy |
| Fiji Women's Crisis Centre | FWCC | 1984 | Mixed | Anti-violence, rape, feminist |
| Fiji Women's Rights Movement | FWRM | 1986 | Mixed | Working, legal and other rights, feminist |
| Women's Action for Change | WAC | 1993 | Mixed | Community theatre, feminist |
| Fiji Association of Garment Workers | FAGW | 1989 | Mixed | Industrial association |
| Fiji Nursing Association | FNA | 1956 | Mixed (Fijian dominant) | Union (has male members) |
| FTUC Women's Wing | – | 1983 | Mixed | Union |
| Women's Employment and Economic Rights | WEER | 1993 | Mixed | Sub-project of FWRM |
| Women in Business | – | 1998 | Mixed | Business, community work |
| Women's Social and Economic Development Programme | WOSED | 1993 | Mixed | Government project, micro-credit, business |

*Source*  Author's notes

## *Multiple axes of women's agency and activism*

Chhachhi and Pittin (1996) have presented a convincing case for the multiple axes of women's agency and activism – a reflection of women's 'messy' and 'multiple realities'. They challenge the notion that women's domestic and repro-ductive roles necessarily weaken their consciousness as workers and as agents. Rather, they argue that 'the very multiplicity of roles and plethora of pressures may provide both the impetus and the necessary networking and organisational structures or base for women to organise' (Chhachhi and Pittin 1996: 120).

This appears to be true for Rukaama Lal, a unionist in Fiji with the multiple roles and links theorised by Chhachhi and Pittin. Since 1990, she has been a vice-president of the Fiji Teachers' Union and founding president of the union's women's wing. Other roles include being a wife, mother of six, and head teacher (since 1980) of a primary school; participation in the Red Cross, other social work, and a local women's club; and being secretary of the Sammlit Women's Association in Lautoka (*Fiji's Daily Post* 1 April 1996; *Fiji Times* 13 February 1997). Women's public negotiation of their roles is often through prolonged participation in community groups and, particularly in Fiji, religious and village organisations. Rukaama strongly advocates basing women's equality and devel-opment in family and community as well as in institutions such as school, unions and the workplace. She proudly identifies as a feminist but, like similar women in Fiji, does not see this as separatist but works with 'women only' and mixed gender groups (her union participation embraces both), as well as ethnically based and multi-ethnic groups.

Women's multiple roles may be a strength of their activism but this can lead to being overcommitted to several causes with ensuing problems of insufficient time or resources. The 'selective mobilisation' (Chhachhi and Pittin 1996: 101) of aspects of women's identities may stifle gender resistance, which in Fiji may mean that traditional religious and cultural activities take precedence over femi-nist activism. This may be because of inability to meet time commitments or a deliberate choice to prioritise aspects of identity. For example, the Women in Politics group faced the challenge of mobilising women along gender rather than the usual political lines during the 1999 state elections.

## Women's strategies in Fiji: transposing agency into action

The kind of strategies deployed by women to negotiate change and foster activism are pertinent to understanding the linkages and transitions from indi-vidual agency to belonging to a series to conscious mobilisation in a group. Women's groups in Fiji have been inventive and flexible but often employed strategies that are simultaneously persistent, patient and subtle. These strategies have challenged tradition and patriarchy but also seek acceptance within Fiji's multicultural communities. As the following discussion reveals, while covert resis-tance has been a common strategy in situations where women are silenced and have little power, there has been considerable overt activism, detailed here through women's strike and union activity, political party participation, feminist groups, other new social movements and business involvement. Strategies include structured and spontaneous direct action, personal interventions, intensive media campaigns, education programmes and facilitating women's economic autonomy. The differing identities of ethnicity, tradition, class, religion and nationalism may at times evoke contradictory and conflicting discourses.

### Indirect resistance

Because women's power and spaces have generally been constricted or their actions ignored, as in the media (*Balance* April–June 1999: 6–7), covert resistance becomes particularly relevant to understanding women's agency. In particular, Fiji has a long history of covert protest within women's work (for example, Lal 1985; Shameem 1990a). Examples of contemporary covert resistance in the garment industry include sickness, absenteeism or lingering over toilet breaks. Low productivity and industrial sabotage, such as rubbing lipstick on clothes or incorrectly assembling garments, have reflected resistance to the labour process (Harrington 1999: 204). While the high labour turnover in the garment industry reflects employer tactics of keeping costs low by dismissing employees who qualify for full pay rates, it can also be interpreted as resistance to an authoritarian work environment (Forsyth 1996: 16).

Indirect agency also indicates the lack of public political space for women in Fiji. Women gained suffrage in 1963, but this hardly ushered in many female politicians, reflecting not only structural discrimination but also gendered cultural controls. Although it has been argued that everyday acts of resistance (Scott 1985) within the domestic sphere can represent agency that forms the seed of later political activism, such analysis may imply a 'false teleology'; at an everyday level, it is questionable whether expressions of resistance such as illness represent a conscious form of agency. Take the case of Sita,[5] an Indo-Fijian mother, who had been well educated at secondary school but whose father refused to sign a consent form for her to train as a nurse. He also forbade her to work as a clerk or train as an air hostess. Sita expressed her inability to achieve occupational independence through anorexia: 'I really cried for the work and still I don't go well now. I didn't eat food for I think about a week.' Instead a marriage was arranged and Sita moved to her husband's family cane farm located in a remote rural area. Subsequent efforts by Sita to pursue paid employment were blocked by her husband, a relatively common occurrence for married Indo-Fijian women in this locality. Sita's anorexia indicates both depression and covert protest, but is this 'a latent and subterranean reservoir of consciousness' that leads on to the transcendence of everyday actions (Cohen 1980: 22)? At a personal level, Sita now expresses agency by ensuring her daughters are educated and not constrained as she was. Sita knows from her own experience that unless there are shifts in cultural expectations of gender roles, education only provides a reservoir for silent protest. Everyday forms of resistance may be expressions of agency but do not necessarily become a challenge to political structures (Akram-Lodhi 1992).

## Overt resistance

Direct agency and activism by women have not found much expression through formal state structures in Fiji. Women there, as throughout the Pacific Islands (Schoeffel Meleisea 1994), have been active in organisations within civil society, ranging from village, community, service and church groups (Schoeffel 1979; Kikau 1986; Thomas 1986; Lukere 1997) to trade unions (Leckie 1999). Although many of these organisations reinforce established political discourse, including gender and ethnic stereotypes, they provide a recognised space where women are publicly active and sometimes challenge political structures. Following Young's (1995) distinction, community groups may reflect women's membership of a series or identify as proactive gendered groups. Trade unions are another important site of women's activism that embrace both passive seriality through membership (along with males) as well as conscious group identification as women workers and along class lines.

### Agency at work: unions

Although the actual percentage of female union members remains relatively low at 22 per cent in Fiji (South Pacific and Oceanic Council of Trade Unions (SPOCTU) 1993), women have actively expressed agency and resistance within organised labour. In 1920, Indo-Fijian women participated in a major public works strike. The media warned of women organised into 'intimidation parties', 'too awful to be at large. Last night they hunted in packs, chasing "boys" into their very homes' (*Fiji Times* 11 February 1920; see also Ali 1980: 48–60). During Fiji's first recorded strike in the garment industry in 1960, women took to the streets and brandished placards proclaiming:

We are workers, not slaves

We will not take any more insults

On strike for our rights

£1/5/- = a week's wage?[6]

(*Fiji Times* 28 July 1960)

This protest was mild compared to strikes staged by female garment workers (Leckie 1992: 9–12), after tax-free export factories were established in post-coup Fiji. Wages were extremely low (hourly rates were often under 50 cents Fijian or US$0.25) while employment conditions frequently were substandard. Initially the military state banned unions, but in 1989 the Fiji Association of Garment Workers gained registration. However, because of state and employer opposition as well as internal problems within the union, it soon became a spent force.

Despite an absence of union representation, women in the manufacturing sector continue to protest against their conditions. For example, Chinese migrant workers went on strike in 1998 over exorbitant fees being charged by labour recruiters in China (*Fiji Times* 12 May 1998). These women locked themselves in their apartments (contesting media statements claiming that they were locked in) and went on a hunger strike. Despite their solidarity with other workers being impeded by language barriers as well as spatial isolation, these workers managed to challenge the constraints common to most women in Fiji's garment industry and they highlight their vulnerability as migrants.

In contrast, banking, nursing and teaching are employment sectors where women have high participation rates and actively negotiate working conditions. These women tend to be well educated, are more articulate and have excellent union coverage compared to their sisters in the garment industry. They have also taken industrial action. For example, women played a pivotal role during a teachers' strike in 1985 and bank strikes during 1996–8 over workplace restructuring and redundancies. Fiji's nurses went on strike in 1990 (Leckie 1997a: 144). However, not only was this action strongly contested by conservative women's organisations, such as the National Council of Women (FNCW), but Fiji Nursing Association (FNA) members themselves were divided and faced personal dilemmas over taking industrial action. Mereani Tukana, FNA's president, looked to her faith to reinforce her agency and activism:

> I was scared but I know I was doing the right thing, we were doing the right thing ... I am a Christian, I sought help from that and when I was given my Scriptures for that week, confirmed to me that yes go ahead, I believe that so that's how I came to the decision.
>
> (Personal interview, 12 September 1996)

Although many nurses publicly joined in with solidarity pickets, some separated this from their domestic lives by concealing their strike participation from their husbands. In some instances, this reflected gender and cultural constraints on involvement in unions and aggressive 'unfeminine' industrial action. As elucidated below, constraints might include prohibitions on public mobility or challenging doctors and politicians who were also traditional Fijian chiefs. This apparent contradiction between overt resistance and private compliance to state–work relations was one way many indigenous Fijian nurses resolved being married to men with conservative political and traditional allegiances. Nurses' industrial agency was also complicated by ethnicity. According to an indigenous unionist (who was president of the Students' Nursing Association in 1983 and has been an FNA council member since 1990), although it appeared that Indo-Fijian women were silent FNA members, this did not apply to comparable institutions. Conversely, indigenous Fijian participation did not necessarily imply active involvement in decision-making.

> We Fijians ... accept whatever we are given and women tend not to speak out too much in our culture ... In the other unions the Indians are very active. This [FNA] is one union that they are not. I have always told my

colleagues if you sit in on the teachers' union you will see Indian women very vocal. If you look in the papers there will be a lot of other Indian organisations; women are very vocal. This is one organisation where they are not very active in, they play silent roles, I don't know, maybe because we dominate them in number, the amount of Fijian nurses working. But then it's not a very good point because the Fijian nurses don't speak up much. You should sit in on one of our branch meetings: most of them, they will attend in number but they will keep quiet. It's only a few people talking and these few people dominate the meeting and if they pull a decision one way and then that's it.

(Personal interview, 25 October 1996)

FNA is an assertive feminised union, but competing identities include those of ethnicity, hierarchy and being a union and/or professional organisation.

Education has been a cornerstone in facilitating women's working rights. In Fiji since the late 1970s, unions have been proactive in programmes which stress gender awareness and issues pertinent to women. This impetus was renewed in 1990 when the South Pacific and Oceanic Council of Trade Unions (SPOCTU) and the International Confederation of Free Trade Unions–Asia Pacific Regional Office (ICFTU–APRO) inaugurated a project to integrate new and established women workers into trade unions in the Pacific Islands (SPOCTU 1993). Padma Dorai, ICFTU–APRO's Director of the Women's Department, assessed that by 1996 gender awareness was still minimal within Pacific unions, and noted that the new ICFTU gender education project imposed tighter but less ambitious strategies because 'change is not to be forced down people's throats' but brought about gradually through promoting gender awareness.[7]

Securing improvements to women's working conditions and pay remains a key focus of collective action within trade unions and feminist organisations in Fiji. Should such advocacy be mainstream or relegated to separate women's caucuses, such as the Women's Wing of the Fiji Trades Union Congress and the women's committees within civil service unions (Leckie 1997b: 62–4)? According to Praveen Sharma, former co-ordinator of FWRM's Women, Employment and Economic Rights Project (WEER), issues pertinent to women workers were increasingly marginalised during post-coup confrontations between unions and the state. After unions were weakened by labour reforms in 1991, FWRM raised its advocacy role for women workers' rights (*Balance* July–August 1997: 4). FWRM lobbied on broader issues than just organised labour, seeking legal recognition for women's unpaid labour within the home and subsistence agriculture (*Balance* April 1997). Even on issues concerning paid work, feminist and union strategies have sometimes diverged, as in the controversy over lifting restrictions on women in paid employment at night. On the one hand, some unionists argued that first there must be improved conditions for all night workers, because if women had open access to paid night work it would widen the scope for their exploitation. Employers, particularly within the garment industry, were one of the pressure groups pushing for a liberalisation of night employment. On the other hand, FWRM and FWCC (Sharma 1996: 1) stressed that women workers such as nurses have always worked at night. Other women, as in the garment industry, have worked illegally on night shifts. Feminists argued for change on the principle of women's right to choose or refuse night work. Arguments against night work based on protecting women and the sanctity of the family were also rejected. The ban on women working at night was lifted in 1996.

*Within the corridors of power: political parties*

Women's rising participation in paid work in Fiji has not been matched by their political representation. Until 1999, Fiji's record of female politicians was dismal (Booth 1994: 2–10; Drage 1998). In 1966, Irene Jai Narayan became the first woman elected to Fiji's Legislative Council representing a mainly Indo-Fijian opposition (Ritova 1996). After the coups she was one of the few Indo-Fijians to serve as a minister within the military government. Adi Finau Tabakaucoro, an ethnic Fijian, headed the newly established Ministry for Women's Affairs. Her ethnic and nationalist sympathies were clarified when she addressed women's groups: 'I do not support any efforts that push women only because we are women at the expense of national politics' (quoted in Emberson-Bain 1992: 156). Tabakaucoro went on to become one of the few female senators in Fiji's upper house. In early 1999, only two women remained within the seventy-member parliament: Taufa Vakatale, Deputy Prime Minister and Minister for Education, and Seruwaia Hong Tiy, Minister for Women's Affairs. The number of women politicians within opposition parties was also small but increased greatly during elections in mid-1999. One prominent female politician is Adi Kuini Speed, a former leader of the multi-ethnic Fiji Labour Party and, in 1998, leader of the indigenous Fiji Association Party. Her title as a chief has aided her acceptance in national politics and ability to cross between social democrat and traditional political spaces (Naulivou 1996). She also commands status as one of five women among the fifty-one *Bose Levu Vakaturaga* (Great Council of Chiefs). Local government has increasingly attracted female candidates, although to date mostly indigenous women have been elected. In 1993, six out of twenty-one women candidates gained local body seats. By 1996, fourteen women, including a mayor, were local body councillors (Drage 1998). Imrana Jalal (1997b: 85) contends that indigenous women have accessed the public realm more success-fully than Indo-Fijian women because of linkages through traditional chiefly power.

Since 1995, Women in Politics in the Pacific Centre (WIPPAC) has been a regional initiative to boost women's political participation. This includes efforts to lobby and monitor politicians so they are accountable to gender issues. The United Nations Development Fund for Women (UNIFEM), has been a catalyst in the operation of WIPPAC. Partly through the initiative of WIPPAC, Fiji's large multi-ethnic NGO, the FNCW sponsors Women in Politics (WIP), a project to boost women's involvement in politics and public organisations, enhance women's skills and understanding of governance and develop a politicised female constituency. This was to counteract feedback from WIP workshops that women voted on racial rather than gender lines (*The Review* July 1998).

## Pressure from feminist groups

WIP may be considered as an organisation where women are constituted as members of a series. This commonality of exclusion from political spaces includes women of radically different political ideologies. Feminist groups in Fiji also embrace women of differing political persuasions, ethnicities and classes but have consciously exhibited assertiveness as social collectives.

The formation of new women's activist groups in Fiji from the 1980s has reflected the dissatisfaction of mounting numbers of women with 'silent resistance' and conventional 'ladies' organisations'. This has followed global shifts where women have become proactive through both feminist movements and within existing organisations, such as trade unions. At a national federation party women's movement convention in 1997, Jalal (cited in *Balance* September–October 1997: 9) articulated the discourse of direct activism, considered by some in Fiji as an affront to cultures there:

> We cannot be considered equals when we are constantly lowering our heads out of respect both in the literal and metaphorical sense. To demand our rights means we have to look at those in power directly in their eyes and demand our rights.

Although tarred as 'home breakers', the FWCC in fact places women's role in the family at the centre of its programmes. Co-ordinator Shamima Ali asserted: 'We are based on the principles of feminism, we empower women, we believe in family' (*Fiji Times* 10 March 1998). A key strategy is to inform women of their rights and then provide support as women make choices.

The newer women's organisations have been confronting the realities of women's agency and choices against Fiji's traditions and economic and political pressures. FWCC and FWRM began as activist groups concerned with specific problems, such as rape, violence against women and exploitation in the garment industry. Although these projects broadened to related gender issues, economic inequalities and human rights, strategies have focused on an issue or a project when lobbying politicians, mounting media campaigns or educating women and men. Both the FWCC and the FWRM have taken strong public monitoring roles and made extensive use of media to highlight government and judicial attitudes towards women. WEER is one campaign that ranges from the specific – sexual harassment, 'Girls Can Do Anything', 'Jobs Have No Gender' (*Fiji Times* 5 March 1998) – to the general – workers' rights, poverty, structural adjustment policies. In relation to another campaign, activist Peni Moore (cited in *Balance* April 1997: 5) pointed out that FWRM used rape as an issue to bring the concept of women's rights to a wide range of women in Fiji:

most of us felt more connected to the violence against women issue than the more abstract concept of 'Women's Rights' ... it was the women partici- pants [in workshops] themselves that would identify their problems – that their oppression was the result of the assault and the reason men got away with rape. In this way, they began to identify with Women's Rights.

A turning point for women's actualisation of their agency is self-awareness that they are oppressed not just as individuals, but that their subordination is a gender issue. This may be either as a series of women in relation to common structures or ideologies, or agency may become realised when a collective focuses on a particular issue.

Consciousness raising has been formalised in educational and training projects. The FWRM and FWCC have produced pamphlets, posters and radio programmes in Fiji's main languages; a video titled *Forceline*;[8] conducted work- shops with schools, health and legal workers and police; and educated women about gender violence, legal rights and empowerment. Programmes reach remote localities such as the island of Taveuni, where FWRM and the Taveuni Women's Support Group held a workshop addressing the problems surrounding marriage, child abuse, domestic violence and rape (*Balance* May–June 1998: 2). The models developed by the FWCC and FWRM to combat violence have been applied in other Pacific societies. In 1993, FWRM began training female paralegals on women and the law and were requested by women's groups in other Pacific Islands to assist in establishing similar programmes (*Balance* April 1997: 2).

Feminist organisations have also taken direct action with street marches, such as 'Reclaim the Night' (for women's safety in public spaces). Street theatre, as enacted by Women's Action for Change, has also had a visible advocacy and educational role.

*New social movements*

The qualitative growth of women's political agency in post-coup Fiji has links with alternative political groups, especially those committed to the restoration of democracy and protection of human rights. These include the Back to Early May Movement and Citizens' Constitutional Forum which, while not specifically gender focused, have relied upon a strong politically aware female input (Griffen 1997). In 1996, the Women's Coalition for Women's Citizenship Rights (WCFWCR) lobbied politicians to change discriminatory aspects of Fiji's constitution. Women's activism has also been linked with other new social movements, particularly the anti-nuclear movements, including the Nuclear Free and Independent Pacific Movement and, locally, the Fiji Anti Nuclear Group (FANG).

*Business and agency*

A less politicised strategy, but one which promotes women's autonomy and indi- vidual agency, has been the facilitation of women in business. Since 1993, the

Women's Social and Economic Development Programme (WOSED) has encouraged women's access to microcredit and entry into small businesses. It is debatable whether this strategy transposes into collective agency (Osmani 1998). Also contentious is whether such gender projects retain any activist or independent identity when they are linked with mainstream structures, such as Fiji's Ministry of Women and Culture (cf. Kabeer 1994: 264–305). At a more practical level, several problems have arisen in establishing microcredit schemes in the Pacific due to small populations, inadequate management, resources, repayments, and support (see Yska 1998: 7). In Fiji, scope for ethnic competition or favouritism for indigenous Fijians has complicated women's access to credit schemes.

Women in Business, a professional organisation established in the late 1990s (*Fiji Times* 25 April 1998) seeks to combine women's business development with other forms of agency. Its aims include offering mutual support, networking and encouraging professional standards and community service, as well as taking an advocacy role in women's education and employment.

## Culture and women's agency in Fiji

The question of cultural specificity and women's agency is fraught with methodological difficulties, notably the politics of representation. Mohanty's (1988) critique of the image of the 'Third World woman' as a passive victim of patriarchy and capitalism is pertinent to representations of Pacific Island women. This is where the notion of seriality is relevant, rather than infinitely breaking multiplicities down to the individual with the ensuing problems for collective agency. However, many of the ethnographic and ethnohistoric accounts as well as much of the political discourse in Fiji stress the specifics of culture in relation to women's agency and political activism. In order to centre women's concerns and identify the context against which women are members of a series or a self-identifying group, the specifics of culture and history require attention.

The ideology of *purdah* is relevant in the lives of many (Lateef 1990b: 44–8) but not all (cf. Mohanty 1988: 66) Indo-Fijian women. This discursively delineates cultural and spatial constraints, segregation of the sexes and the reinforcement of men's protection of women and family. Lateef (1992: 201) has emphasised how the family creates divisions and competition between women rather than fostering solidarity. Carswell's recent fieldwork in a cane-farming district found Indo-Fijian women's agency was heavily circumscribed by community values (refer to note 4). Women's status was based on being chaste, virtuous, well mannered and respected – that is, fulfilling, *dharma* (Kelly 1991: 236–7). These women could participate in religious or fund-raising groups but faced limitations on such voluntary work, limitations which included attaining their husband's consent to be mobile outside the home and the censure of gossip against a woman who was seen 'rushing here and there' (personal communication, Sue Carswell, August 1997).

Many accounts of gender relations within indigenous communities also stress men's power and authority over women. This derives partly from men's greater access to resources, especially land, which in turn is a basis of power and respect and confers titles and political power within both traditional and state politics:

> Men exercise authority over women. Their authority derives from a number of sources. First, the fact that they are leaders at the societal level – the district, village and clan heads are always male – justifies extension of 'male authority' over family members. In addition, men have control over religious practices that centre on the clans' men's houses (*na beto*). Finally, they are considered stronger than women, and strength is valued in Fijian society. These factors allow them to lead society and to be leaders within the clan and the household.
>
> (McKenzie Aucoin 1990: 26; see also Ravuvu 1987: 261–80)

Men's power has been bolstered by customary hierarchies, patrilineal inheritance of titles and land, and patrilocal residence of women, which curbs their formalised land rights (Jalal 1998: 53–7). 'Men hold and control leadership positions and are said to be the owners (*na leya*) of the clans, villages and districts' (McKenzie Aucoin 1990: 26).

Colonial and post-colonial state policy strengthened patrilineal transfers of land, formalising 'a patrilineal and patriarchal system of land management and control' (Jalal 1997b: 83). This was despite regional variation of gender relations between Fijian communities, including matrilineal patterns in some areas. Customary law determines women's limited control over land and any rights they have are only considered as usufruct (Jalal 1997b: 84). Likewise, chiefly titles are usually inherited through the male line. These customs could be challenged under Fiji's 1997 constitution but there is likely to be strong resistance to such change.

It could be argued that men's power was strengthened by colonial structures and discourse which reinforced chiefly hierarchies and patriarchy. The colonial government dealt mostly with men, endorsing a 'male perspective of custom, a reflection of colonial stereotypes about women' (Jalal 1997b: 83). This was underscored by Christian teachings of men's authority over women, and women's restriction to a private domestic sphere (Ralston 1990; Schoeffel Meleisea 1994: 111; Lukere 1997).

Cultural specifics are significant when addressing women's agency in Fiji but there are also gender issues that cut through ethnicity/class/region. This is not to prioritise gender as an absolute or to marginalise the dominance of ethnicity/class/tradition in Fiji's politics. But, as argued by many women's organisations in Fiji, women face a common gendered culture. Jalal and former parliamentarian Wadan Narsey have termed this a 'culture of silence', condemning any assertiveness by women as being disrespectful to those with traditional power (*Fiji Times* 19 June 1997). Rukaama Lal has also questioned this discourse of gender inequalities: 'While men dominate the household, women are left behind, are subservient and often abused and overworked. This is partially accepted because of religious and cultural teachings, but how long are we going to continue to take this?' (*Fiji Times* 13 February 1997).

Women leaders in Fiji have begun to caution women against identifying as victims or being self-deprecating and 'convinced of their own unfitness' (in the words of former Women and Culture Minister Seruwaia Hong Tiy, *Fiji Times* 20 April 1998). A critical shift from agency to activism is considered to occur when a woman decides to act if options to facilitate this are present. Even so, many women resolve the dilemma between awareness and action by remaining silent:

> Many women in fact *do know* what they want to do and also what they think would be the best thing to do in what often are painful and difficult situations. But there seems to be that fear that others would condemn or hurt them if they speak, that others would not listen or understand, that speaking would only lead to further confusion, that it was better to appear selfless, to give up their voices and keep the peace.
>
> (Jalal, cited in *Fiji Times* 19 June 1997)

Women's political organisations can be crucial gateways to an individual's awareness of the politics of her insubordination and ability to challenge this, either in belonging to a series or in a more politically conscious and cohesive way. Equally, women call upon supportive family, friends, church and other community networks which act as further spaces for the contradictions of group activism and specific identity.

## The post-coup milieu and women's agency

If there are both specifics of gender ideology within cultures and a gendered cross-ethnic, cross-class culture in Fiji, then what has been the impact of developments in Fiji since the late 1980s? What meaning did the coups and the post-coup regimes have for women's agency and activism? These regimes constituted part of the milieu that framed women's seriality, yet through resistance to post-coup conditions new politicised and cohesive groups emerged.

Initially the coup was condemned as highly negative for women's rights among all ethnicities. The impact was especially severe for Indo-Fijian women. According to Lateef (1990a: 121), 'The coups represent a retrogressive step in their struggle for greater freedom.' They suffered the double bind of racial and gender discrimination, exacerbated by worsening economic conditions (Knapman 1988). The coups accentuated a chilling fear of ethnic violence which greatly restricted Indo-Fijian women's spatial mobility. Some Indo-Fijian communities reacted to ethnic discrimination by reasserting identity through a conservative strengthening of tradition and patriarchy (cf. Charles and Hintjens 1998).

Indo-Fijian women's groups also suffered in the competition for scarce resources during the post-coup economic crisis and economic recession of the late 1990s. For example, Ra Nari Kaliyan Parishad, an Indo-Fijian association of rural-based women's clubs, languished because of inadequate funding from government or donors (Subramani 1997: 28). In the immediate post-coup years the state favoured indigenous and mainstream organisations, such as the FNCW. A study found that although 'the crucial aspect of NGOs is their very non-conformity to government values and accepted norms' (Singh 1994: 204), in practice 'the ethnicity of the people involved in running the organisation is supported by the government – and certain donors by extension. This has serious implications for the independence of NGOs' (Singh 1994: 199, cited in Subramani 1997: 28). For example, the Soqosoqo Vakamarama in effect became the women's wing of the Soqosoqo Vakavulewa Ni Taukei (SVT) political party which formed the government between 1992 and 1999.

Some women's groups met with more direct coercion from the government. For example, on 8 March 1996, a permit for women to march through Labasa to commemorate International Women's Day was revoked. The march had been organised by mainly Indo-Fijians belonging to the Democratic Women's Forum (Prajatantiya Mahila Manch), formed in 1990. Instead, the government organised a separate programme at the army grounds (*Sangarsh* 9 March 1996).

Indigenous women's agency has also been restrained by the post-coup revival of tradition. 'Identity politics promise women security and meaning in what, from a conservative point of view, is a world gone mad. Stability lies in part in clearly defined sex roles, family life, and a religious orientation' (Moghadam 1994: 19). However, the relevance of what constitutes tradition is contested among Fijian women. The contradictions of gender equality and the politics of tradition emerged over Fiji's ratification of the United Nations Convention on the Elimination of All Forms of Discrimination Against Women (CEDAW). Fiji entered a reservation, based on cultural and traditional values, to two CEDAW articles.[9] These objections have been superseded by Fiji's 1997 constitution.

Another contentious issue was 'gender tokenism', when select indigenous women were offered appointments during the years of military rule. Chiefly rank and state political connections have also controlled women's post-coup activism (Leckie 1997a: 141–2).

## Conclusion: agency and difference

Women's political participation accelerated when they comprised 22 per cent of candidates in the 1999 elections and gained eight parliamentary seats. Under a new 'People's Coalition', five women had ministerial portfolios and Adi Kuini Speed was a deputy prime minister. The new government was committed to ameliorating discrimination against women, while other possibilities such as a minimum wage could have improved women's economic status. The ascent of women's presence in high-level politics, including the appointment of three women activists to Senate,[10] was the result of differing paths of women's agency in Fiji. In the build-up to the elections the Fiji Women's Political Caucus (a WIP initiative) endorsed gendered political agency. This example of gendered seriality embraced women from different parties and ethnicities, providing practical support, including access to an election fund. Women candidates called upon women voters to vote for females rather than along party lines (PACNEWS 9

March 1999). A new preferential voting system in Fiji provided the structural option of being able to select candidates on both party and gender lines.

This chapter has argued that there is a specific women's agency, but it is complicated with other identities. The cultural constraints discussed in this paper are negotiated but continue to be reproduced in contemporary Fiji. This reinforces the differences among women and can fracture activism. Economic circumstances also exert powerful unifying and dividing influences on women. Levels of poverty in Fiji are increasing among all ethnic groups but women are particularly vulnerable (UNDP/Government of Fiji 1997: 53–4). Market-based solutions to Fiji's development crises, whether directed at promoting export manufacturing or the informal sector (UNDP 1997), especially affect women. Fiji's economy also depends upon tourism, where women's work as well as their exotic images are crucial (Jolly 1997). The sugar industry, the third prong of the export economy, depends upon women's unpaid and unrecognised labour. Women are not always compliant workers but there are considerable constraints on their unionisation in many sectors. Labour legislation in 1991 tightened state control over the functioning of unions and industrial action, while also impeding the unionisation of unorganised workers. This especially hampers women who are likely to be cautious about resistance when their employment is insecure. However, some women, such as nurses, who are in secure employment may be wary of being assertive, as this could count against career advancement. Even when compliant, women face discrimination over promotions to top-level public and private positions (Booth 1994: 8; Forsyth 1996). This lack of representation at top levels extends to other sectors of public life (Samisoni 1998). Cultural specifics are significant here, but the gendered culture of patriarchy is pertinent to most women in Fiji. FWRM's television campaign against sexual harassment provides an example of the contradictions of gender and ethnic representation. Objections, mainly from men, have been made to advertisements where the victim is portrayed as an indigenous female, the harasser as an Indo-Fijian male work supervisor and the manager as a European. As Jalal (1997a) has argued, while this paints an ugly picture of Indo-Fijian males, had FWRM chosen indigenous Fijians either in the dominant roles or for all the subjects it would have been accused of being anti-Fijian. Objections to the representation distracted from the 'real issues' of gender harassment. 'Is the racism so deeply embedded that we cannot fathom the discrimination faced by Fiji women because we are so blinded by racial issues?' (Jalal 1997a).

Young's (1995) theorisation of gender as seriality is salient to women's activism in Fiji because women do identify within the common category 'woman' but also align with ethnic, religious and traditional identities. Feminist and mainstream political groups, as well as organisations like unions and, to a much lesser extent, business and professional networks, mobilise according to consciousness of common constraints as women, but often this can at best only be a series due to the continued prominence of ethnicity and tradition (such as in FNA) and fractures along class lines. Traditional women's groups, notably the Soqosoqo Vakamarama, the Methodist Women's Fellowship and a multitude of

Indo-Fijian organisations, continue to appeal to their conventional constituencies but are connecting with women from the FNCW, FWCC, FWRM and other new social movements through campaigns against violence, human rights abuses or advocating women's participation in politics, the workforce and business. These networks remain loosely affiliated partly because of the reassertion of ethnic politics after the 1987 coups, but also because of the differential impact of economic restructuring not only between men and women but among women.

Yet women's agency and activism in Fiji cannot just be reduced to 'series of convenience' that arise, for example, before an election or following publicity of horrific violence against women. The FWRM and FWCC are overtly conscious social collectives. Although the coups sharpened racialism which had dire consequences for women's collective agency, this 'protracted crisis' was the beginning of a transformative period for civil society (Subramani 1997: 30). There has been a measurable growth in gender politics and a realisation of gender as well as ethnic and class subordination. Gender concerns are central to human rights and, in turn, the recognition of human rights and difference has enhanced support for women's causes in Fiji. While women's issues are receiving greater recognition both within mainstream and alternative institutions, the basis of agency rests in the everyday concerns of women from all ethnicities and classes in Fiji, and in the creative solutions they employ – sometimes through gender seriality and other times through the complexities of women's more specific constituencies.

## Acknowledgements

This chapter arose from a team project on *Work, Identity, Ethnicity and Gender in Fiji*, based around fieldwork in the sugar and garments industries, subsistence and public sectors in Fiji. Many of my views are based on personal experience working as a teacher, researcher and activist in Fiji since 1982. In this paper I have welcomed insights and assistance from Sue Carswell, Christy Harrington, Graham Boyle, Claire Slatter, Ruth Schick, Tupou Vere (Fiji Ministry of Women and Culture), Fiji Nursing Association, Fiji Women's Rights Movement and many anonymous informants. The University of Otago generously provided funding.

## Notes

1 This refers to the hierarchical nature of indigenous Fijian society and the dominance of chiefs. This high political and social status is derived from patrilineal agnatic descent. Male chiefs usually have the title *ratu* and female chiefs *adi*. Chiefs lead the *yavusa* or 'clan' and smaller lineages within this of *mataqali* or land-owning and descent groups. This somewhat idealised structure was codified under colonialism. In fact there was and is considerable customary diversity in Fiji into the significance of chiefly rank, the structure of social hierarchies and the dynamics between chiefs and commoners. Note that the term 'chiefly system' is commonly used in Fiji and refers to the dominance and centrality of people known as chiefs in this society, as well as indicating values of respect, deference, adherence to tradition, etc.

2 This chapter was written before 19 May 2000 when the government elected in 1999 was overthrown by another coup. Parliamentary elections were held in

August–September 2001 with a conservative pro-Fijian party being elected. These results are currently being challenged in court by the Fiji Labour Party.

3 Phyllis Herda is editing such a collection in *Portraits of Polynesian Women.*

4 Statement by Adi Ema Tagicakibau, an indigenous Fijian organiser in FWRM. She was elected to parliament in 1999 and was Assistant Minister to the Prime Minister.

5 Her story was recounted to me by Sue Carswell who was completing a PhD thesis on women and children in Fiji's sugar cane industry (Anthropology Department, University of Otago) at the time this article was being written.

6 Approximately US$3.00.

7 Presentation at the *FTUC/ICFTU–APRO Seminar on Gender Perspectives*, 2 September 1996, Suva.

8 'Forceline' is a local term implying that rape is an acceptable cultural practice.

9 These are: (i) Article 5(a) which aims to achieve the elimination of prejudices, customary and all other practices which are based on the idea of the inferiority or the superiority of either of the sexes or on stereotyped roles for men and women; and (ii) Article 9 covering the equal rights of nationality and the nationality of children.

10 Atu Emberson-Bain, Jokapeci Koroi (President of the Fiji Labour Party and former leader of the FNA) and Usha Devi Singh.

# References

Akram-Lodhi, A.H. (1992) 'Peasants and hegemony in the work of James C. Scott', *Peasant Studies* 19, 3/4: 179–201.

Ali, A. (1980) *Plantation to Politics: Studies on Fiji Indians*, Suva: University of the South Pacific, *Fiji Times* and the *Herald.*

*Balance: Newsletter of the Fiji Women's Rights Movement* (1996–9), Suva.

Bhabha, H. (1994) *The Location of Culture*, London: Routledge.

Booth, H. (1994) *Women of Fiji: A Statistical Gender Profile*, Suva: Department for Women and Culture.

Brunt, R. (1989) 'The politics of identity', in S. Hall and M. Jacques (eds) *New Times: The Changing Face of Politics in the 1990s*, London: Lawrence and Wishart.

Charles, N. and Hintjens, H. (1998) 'Gender, ethnicity and cultural identity: women's "places"', in N. Charles and H. Hintjens (eds) *Gender, Ethnicity and Political Ideologies*, London: Routledge.

Chhachhi, A. and Pittin, R. (1996) 'Multiple identities, multiple strategies', in A. Chhachhi and R. Pittin (eds) *Confronting State, Capital and Patriarchy: Women Organizing in the Process of Industrialization*, New York: St Martin's Press.

Cohen, R. (1980) 'Resistance and hidden forms of consciousness among African workers', *Review of African Political Economy*, 19: 8–22.

Drage, J. (1998) 'Women and politics in the Pacific', paper presented at the *6th Pacific Islands Political Studies Association Conference*, University of Canterbury, Christchurch.

Emberson-Bain, A. (1992) 'Women, poverty and post-coup pressure', in D. Robie (ed.) *Tu Galala: Social Change in the South Pacific*, Wellington: Bridget Williams/Pluto.

*Fiji Times*, Suva, various issues.

*Fiji's Daily Post*, Suva, various issues.

Forsyth, D. (1996) 'Women workers in Fiji's formal sector', unpublished report, University of the South Pacific, Suva.

Geraghty, P. (1997) 'The ethnic basis of society', in B. Lal and T. Vakatora (eds) *Fiji in Transition: Fiji Constitutional Review Commission*, Research Papers 1, Suva: School of Social and Economic Development, University of the South Pacific.

Griffen, A. (ed.) (1997) *With Heart and Nerve and Sinew. Post-coup Writing From Fiji*, Suva: Christmas Club.

Haggis, J. (1998) ' "Good wives and mothers" or "dedicated workers"? Contradictions of domesticity in the "mission of sisterhood", Tranvancore, South India', in K. Ram and M. Jolly (eds) *Maternities and Modernities: Colonial and Postcolonial Experiences in Asia and the Pacific*, Cambridge: Cambridge University Press.

Harrington, C. (1999), 'The seams of subjectivity and structure: women's experiences of garment work in Aotearoa New Zealand and Fiji', unpublished PhD thesis, University of Otago, New Zealand.

Jalal, P.I. (1997a) 'Race clouds gender issue', *Fiji Times*, Suva, 27 March.

——(1997b) 'The status of Fiji women and the constitution', in B. Lal and T. Vakatora (eds) *Fiji in Transition: Fiji Constitutional Review Commission*, Research Papers 1, Suva: School of Social and Economic Development, University of the South Pacific.

——(1998) *Law for Pacific Women. A Legal Rights Handbook*, Suva: Fiji Women's Rights Movement.

Jolly, M. (1997) 'From Point Venus to Bali H'ai: eroticism and exoticism in representations of the Pacific', in L. Manderson and M. Jolly (eds) *Sites of Desire? Economies of Pleasure: Sexualities in Asia and the Pacific*, Chicago: Chicago University Press.

——(1998a) 'Introduction: colonial and postcolonial plots in histories of maternities and modernities', in K. Ram and M. Jolly (eds) *Maternities and Modernities: Colonial and Post-colonial Experiences in Asia and the Pacific*, Cambridge: Cambridge University Press.

——(1998b) 'Other mothers: maternal "insouciance" and the depopulation debate in Fiji and Vanuatu, 1890–1930', in K. Ram and M. Jolly (eds) *Maternities and Modernities: Colonial and Postcolonial Experiences in Asia and the Pacific*, Cambridge: Cambridge University Press.

Kabeer, N. (1994) *Reversed Realities: Gender Hierarchies in Development Thought*, London: Verso.

Kelly, J.D. (1991) *A Politics of Virtue: Hinduism, Sexuality and Countercolonial Discourse in Fiji*, Chicago: University of Chicago Press.

Kikau, E. (1986) 'Of challenges and choices: women's organisations in rural development', unpublished MA thesis, University of Reading, UK.

Knapman, B. (1988) 'Afterword: the economic consequences of the coups', in R. Robertson and A. Tamanisau (eds) *Fiji: Shattered Coups*, Sydney: Pluto Press.

Lal, B.V. (1985) 'Kunti's cry: indentured women in Fiji plantations', *Indian Economic and Social History Review* 22, 1: 55–71.

——(1992) *Broken Waves: A History of the Fiji Islands in the Twentieth Century*, Honolulu: University of Hawai'i Press.

Lateef, S. (1990a) 'Current and future implications of the coups for women in Fiji', *Contemporary Pacific* 2, 1: 113–30.

——(1990b) 'Rule by the *danda*: domestic violence among Indo-Fijians', *Pacific Studies* 13, 3: 43–62.

——(1992) 'Wife abuse among Indo-Fijians', in D.A. Counts, J.K. Brown and J.C. Campbell (eds) *Sanctions and Sanctuary: Cultural Perspectives on the Beating of Wives*, Boulder: Westview Press.

Leckie, J. (1990) 'Workers in colonial Fiji', in C. Moore, J. Leckie and D. Munro (eds) *Labour in the South Pacific*, Townsville: James Cook University of North Queensland.

——(1992) 'Industrial relations in post-coup Fiji: a taste of the 1990s', *New Zealand Journal of Industrial Relations* 17: 5–21.

——(1997a) 'Gender and work in Fiji: constraints to re-negotiation', *Women's Studies Journal* 13, 2: 127–53.

——(1997b) *To Labour with the State: The Fiji Public Service Association*, Dunedin: University of Otago Press.

——(1999) 'Silence at work: trade unions and gender diversity in the South Pacific', in G. Hunt (ed.) *Labouring for Rights: An International Assessment of Union Response to Sexual Diversity*, Philadelphia: Temple University Press.

Linnekin, J. (1990) *Sacred Queens and Women of Consequence. Rank, Gender and Colonialism in the Hawaiian Islands*, Ann Arbor: University of Michigan Press.

Lukere, V. (1997) 'Mothers of the Taukei: Fijian women and the decrease of the race', unpublished PhD thesis, Australian National University, Canberra.

Matahaere-Atariki, D. (1998) 'At the gates of the knowledge factory: voice, authenticity, and the limits of representation', in R. Du Plessis and L. Alice (eds) *Feminist Thought in Aotearoa/New Zealand. Connections and Differences*, Auckland: Oxford University Press.

McKenzie Aucoin, P. (1990) 'Domestic violence and social relations of conflict in Fiji', *Pacific Studies* 13, 3: 23–42.

Moghadam, V. (1994) 'Introduction: women and identity politics in theoretical and comparative perspective', in V.M. Moghadam (ed.) *Identity Politics and Women. Cultural Reassertions and Feminisms in International Perspective*, Boulder: Westview Press.

Mohanty, C.T. (1988) 'Under western eyes: feminist scholarship and colonial discourses', *Feminist Review* 30: 61–88.

Mouffe, C. (1993) 'Feminism, citizenship and radical democratic politics', in C. Mouffe, *The Return of the Political*, London: Verso.

Naulivou, S. (1996) 'Adi Kuini reflects on her childhood days', *Marama Vou*, November: 22–3.

Nicholson, L. (1995) 'Interpreting gender', in L. Nicholson and S. Seidman (eds) *Social Postmodernism: Beyond Identity Politics*, Cambridge: Cambridge University Press.

Nicholson, L. and Seidman, S. (1995) 'Introduction', in L. Nicholson and S. Seidman (eds) *Social Postmodernism: Beyond Identity Politics*, Cambridge: Cambridge University Press.

Osmani, L.N.K. (1998) 'The Grameen Bank experiment: empowerment of women through credit', in H. Afshar (ed.) *Women and Empowerment. Illustrations from the Third World*, Houndmills: Macmillan.

PACNEWS (1999) Available online http://pidp.ewc.hawaii.edu/pireport (9 March 1999).

Ralston, C. (1990) 'Women workers in Samoa and Tonga in the early twentieth century', in C. Moore, J. Leckie and D. Munro (eds) *Labour in the South Pacific*, Townsville: James Cook University Press.

——(1992) 'The study of women in the Pacific', *Contemporary Pacific* 3: 162–75.

Ravuvu, A. (1987) *The Fijian Ethos*, Suva: Institute of Pacific Studies.

*Review* (monthly), Suva, various issues.

Ritova, S. (1996) 'Irene Jai Narayan … the iron lady …', *Marama Vou*, November 1996: 14–15.

Robertson, R.T. (1998) *Multiculturalism and Reconciliation in an Indulgent Republic. Fiji after the Coups: 1987–1998*, Suva: Fiji Institute of Applied Studies.

Samisoni, M. (1998) 'Balancing gender in decision making', in *Presentations at the National Congress on Women's Plan of Action 1999–2000*, Suva: Ministry of Women and Culture.

*Sangarsh* (1996) Online posting. Available e-mail: pacific-islands-lcommat;coombs.anu.edu.au (9 March 1996).

Schoeffel, P. (1979) 'Daughters of Sina: a study of gender, status and power in Western Samoa', unpublished PhD thesis, Australian National University, Canberra.

Schoeffel Meleisea, P. (1994) 'Women and political leadership in the Pacific Islands', in C. Daley and M. Nolan (eds) *Suffrage and Beyond. International Feminist Perspectives*, Auckland/Annandale: Auckland University Press/Pluto Press.

Scott, J.C. (1985) *Weapons of the Weak: Everyday Forms of Peasant Resistance*, New Haven: Yale University Press.

Shameem, S. (1990a) 'Girmitiya women in Fiji: work, resistance and survival', in C. Moore, J. Leckie and D. Munro (eds) *Labour in the South Pacific*, Townsville: James Cook University Press.

——(1990b), 'Sugar and spice: wealth accumulation and the labour of Indian women in Fiji, 1879–1930', unpublished PhD thesis, University of Canterbury, Christchurch.

Sharma, P. (1996) 'Lifting the night work ban', *Balance* May–June: 1–2.

Singh, P. (1994) 'Collaboration or constraint? NGOs and donor agency in Fiji's development process', unpublished MA thesis, University of the South Pacific, Suva.

Slatter, C. (1987) 'Women factory workers in Fiji: the "half a loaf" syndrome', *Journal of Pacific Studies* 13: 47–59.

Smith, J. (1994) 'The creation of the world we know: the world-economy and the re-creation of gendered identities', in V.M. Moghadam (ed.) *Identity Politics and Women. Cultural Reassertions and Feminisms in International Perspective*, Boulder: Westview Press.

South Pacific and Oceanic Council of Trade Unions (SPOCTU) (1993) 'Survey of trade union development in the Pacific', unpublished report, Brisbane.

Subramani (1997) 'Civil society: people's participation and building a political consensus in Fiji', in B. Lal and T. Vakatora (eds) *Fiji in Transition: Fiji Constitutional Review Commission*, Research Papers 1, Suva: School of Social and Economic Development, University of the South Pacific.

Sutherland, W. (1992) *Beyond the Politics of Race: An Alternative History of Fiji to 1992*, Canberra: Department of Political and Social Change, Research School of Pacific Studies, Australian National University.

Thomas, P. (1986) 'Dimensions of diffusion: delivering primary healthcare and nutritional information in Western Samoa', unpublished PhD thesis, Australian National University, Canberra.

United Nations Development Programme (UNDP) (1997) *Sustaining Livelihoods. Promoting Informal Sector Growth in Pacific Island Countries*, Suva: United Nations.

United Nations Development Programme (UNDP)/Government of Fiji (1997) *Fiji Poverty Report*, Suva: United Nations/Government of Fiji.

Wilkes, C. (1845) *Narrative of the United States Exploring Expedition during the Years 1838, 1839, 1840, 1841, 1842*, Philadelphia: Lea and Blanchard.

Yeatman, A. (1993) 'Voice and representation in the politics of difference', in S. Gunew and A. Yeatman (eds) *Feminism and the Politics of Difference*, Allen and Unwin/Bridget Williams: St Leonards/Wellington.

Young, I.M. (1995) 'Gender as seriality: thinking about women as a social collective', in L. Nicholson and S. Seidman (eds) *Social Postmodernism: Beyond Identity Politics*, Cambridge: Cambridge University Press.

Yska, G. (1998) 'Micro-enterprise development', in *Presentations at the National Congress on Women's Plan of Action 1999–2000*, Suva: Ministry of Women and Culture, Government of Fiji.

Yuval-Davis, N. (1994) 'Identity politics and women's ethnicity', in V.M. Moghadam (ed.) *Identity Politics and Women. Cultural Reassertions and Feminisms in International Perspective*, Boulder: Westview Press.

# 10 'Asia' in everyday life

## Dealing with difference in contemporary Japan

*Vera Mackie*

It has been estimated that perhaps thirteen million people are currently working as labour migrants in the Asian region (Matsui 1999: 46–7). Since the 1980s, Japan has been one of the receiving countries for migrant labour from Southeast Asia, South Asia and the Middle East. As Japan deals with its place at the centre of regional labour markets, as a magnet drawing immigrant workers from the Asian region and beyond, questions of difference have emerged into public discussion. Until recently, questions of difference were kept at a conceptual distance, safely externalised, seen as happening in a distant past or outside the borders of the Japanese nation-state. With the increased presence of immigrant labour in Japan, such distance has become harder and harder to maintain. In this chapter, I will consider some feminist attempts to deal with difference in the particular context of labour migration to Japan. This discussion will be organised around some themes which I shall call the 'histories of difference', 'spaces of difference', 'everyday practices of difference', 'embodiment of difference' and 'dealing with difference'.

Commentators on current relationships between Japan and other Asian countries must often operate in two parallel time frames: the frame of colonial modernity and the frame of post-coloniality (cf. Barlow 1997). Some contemporary issues are a legacy of the imperial and colonial past: the existence of a substantial minority of the descendants of immigrants from the colonial period, and the claims for compensation by those subject to enforced labour in the Second World War. Other issues reflect Japan's place in the post-war political economy of East Asia: the consumption of products produced in the transnational factories of the Asian region; the gendered effects of tourism on the economies of the region; the issue of the rights of immigrant workers within the boundaries of the Japanese nation-state; and questions of the gendered relationships between military personnel and local people due to the existence of United States military bases on Japanese soil. I would argue, with Gupta and Ferguson (1992: 8), that current perceptions of cultural difference have been historically produced in a world of 'culturally, socially, and economically interconnected spaces'.

Japan, after all, is a metropolitan power, shaped by a history of imperialism and colonialism, in ways which are only starting to be analysed. In this context, some commentators have criticised those feminists in Japan who have failed to challenge taboos against discussion of the imperialist and colonialist past, who continue to speak from a position of ethnocentric authority, and who speak from a position which supposes a unitary identity as the basis for a unified feminism which can speak for all Japanese women (Jung *et al.* 1995; Ryang 1998a). This chapter, however, will focus on the writings and works of those women in Japan who *have* attempted to forge links across perceived national and ethnic boundaries in the particular context of dealing with the situation of immigrant workers in contemporary Japan. These are women who have taken the responsibility of studying their own history in order to earn the right to criticise their present. Their project is compatible with that described by Gayatri Chakravorty Spivak (1990: 62–3):

> I say that you have to take a certain risk: to say 'I won't criticize' is salving your conscience, and allowing you not to do any homework. On the other hand, if you criticize having earned the right to do so, then you are indeed taking a risk and will probably be made welcome, and can hope to be judged with respect.

## Histories of difference

Immigration and emigration have played a more significant role in Japan's modern history than is often recognised. By 1940, there was a total of around half a million Japanese emigrants living in the Americas. Japanese emigrated to the colonies of Taiwan in the late nineteenth century and Korea from 1910. By 1937, 1.8 million Japanese emigrants lived in the puppet state of Manchukuo in north-eastern China, with further movements taking place up to 1945. Hundreds of thousands of Koreans travelled to Japan between 1910 and 1945. To some extent these labour flows were reversed at the end of the Second World War. Japanese emigrants and soldiers were repatriated from Asia and the Pacific, and some Korean residents returned to the Korean peninsula until the Korean War made this difficult. There was some Japanese emigration to the Americas in the 1940s and 1950s, and many Koreans and Taiwanese stayed on in Japan as second- and third-generation communities.

In the 1970s, labour migration to Japan largely took the form of small numbers of skilled migrants from First World countries. New trends emerged in the 1980s with workers from Southeast Asia entering Japan to take up jobs in the service sector, manufacturing, and the construction industry. The rising numbers of illegal immigrant workers in the last two decades reflect the rapid growth of the Japanese economy in the mid- to late 1980s, the rapid appreciation of the yen from 1985, and the disparity between wage levels in Japan and other countries in the region. The high proportion of women from Southeast Asia suggests a connection between particular constructions of gender and ethnicity in the Japanese context, but also reflects the contemporary situation in Thailand and the Philippines. Industrialisation and agrarian transformation have determined young women's migration from rural areas to cities or overseas, while many local economies have become dependent on the remittances of emigrants (Piquero-Ballescas 1996: 30–4; Singhanetra-Renard 1996). An initial pattern of feminised migration to the service sector has shifted to include significant numbers of male workers in construction and manufacturing (Stalker 1994: 248–9). This migration has not been halted by the recession in the Japanese economy in the early 1990s, or the Asian economic crisis of the late 1990s (Athukorala and Manning 1999: 3–4). Despite the relatively small number of workers involved, immigrant labour has become an important structural feature of construction, manufacturing and service industries in Japan (Athukorala and Manning 1999: 27–56).

Particular regions provide different proportions of male or female workers, with quite distinctive employment patterns for male and female immigrants. If we look at alien registration statistics,[1] those countries with striking gender imbalance in favour of males have been Bangladesh, Iraq and Pakistan.[2] Females are disproportionately represented in statistics from the Philippines and Thailand. Similar patterns appear in statistics on illegal immigrants (Kasama 1996: 165–6). Among legal workers by residence status, 'entertainers' are the largest single category. While entertainers enter Japan under a legal visa category, this status often masks employment in hostess bars or massage parlours. Illegal male workers are overwhelmingly employed in construction and factory labour – those jobs described as 'dirty, difficult and dangerous' (*kitanai, kitsui, kiken*) – while the largest categories for illegal immigrant women are bar hostess, factory work, prostitution, dishwashing and waitressing (Kasama 1996: 168). The largest suppliers of immigrant workers by nationality are South Korea, Thailand, China, the Philippines and Malaysia. The year 1988 provided a turning point: there were significant increases in the total number of illegal immigrant workers, with males outnumbering females for the first time, and illegal immigrants entering from a wider range of countries (Itô 1992: 294).

Recent attempts to deal with difference have a history which can be traced back to the 1960s. In Japan, as in the other post-war liberal democracies, new forms of left-wing political activism, often rejecting the methods and emphases of the mainstream Communist and Socialist parties and labour unions, emerged in the 1960s. In Japan, such activism was forged in the demonstrations against the renewal of the US–Japan Security Treaty (*Nichi-Bei Ampo Jôyaku*, or *Ampo*) in 1960 and 1970, and radical student activism and counter-cultural activity in the 1970s. Another issue which reminded people in Japan of their links with other Asian countries was the Vietnam War. Although the post-war Japanese constitution explicitly renounces the right to belligerence, Japan has played an important role in US military strategy in East Asia through its provision of bases and support facilities under the terms of the San Francisco Peace Treaty and the US–Japan Security Treaty of 1952. Japan's 'Self-Defence Force' also has one of the world's largest military budgets. As New Left activists in Japan joined international protests against the Vietnam War, they also focused on the complicity of their own government (Sakamoto 1996).

Women demonstrated alongside their male comrades, but often came to feel dissatisfied with their marginal role in these organisations. While some moved on to form 'women's liberation' groups (Tanaka 1995), others attempted to think through the relationship between the issues surrounding the US–Japan Security Treaty, broader questions of the nature of the post-war Japanese polity, and different forms of discrimination based on gender, class, caste and ethnicity (Kitazawa *et al.* 1996: 27–8). In other words, they placed themselves in the larger context of relationships between nations in the East Asian region.

Some members of these organisations were also led into historical research. They investigated such issues as enforced military prostitution in the Second World War (*Ajia to Josei Kaihô* 1983). Other women attempted to close the temporal gap which separated them from their mothers, who had experienced the militarised state of the 1930s, the age of imperialism and colonialism in Asia, and the Pacific War. They tried to understand how the everyday lives of women gradually came to be imbricated in state policies. Their research revealed an intimate connection between state policies and the everyday practices of reproduction and childcare, food preparation and thrift campaigns, dress, and the rituals of everyday life (Kanô 1995). Their research on women's complicity with the militarist state led them to think about the workings of the post-war Japanese political system and the roles women should play in this system. One group, devoted to writing the 'History of the Homefront', called themselves 'Women Questioning the Present' (*Josei no Ima o Tou Kai*), signalling that their research was ultimately concerned with writing a history of their own present situation.

Relations between 'majority' Japanese and Okinawan, Ainu, Korean and Chinese minorities are a constant reminder of Japan's colonial history. A generation of historians in Japan has been engaged in the project of recording the histories of former enforced labourers, war orphans, internees and survivors of the military prostitution system. Although they have not, until recently, labelled this project as 'post-colonial', their writings can usefully be brought into dialogue with critiques of post-coloniality in other parts of the world (cf. Spivak 1987, 1990, 1993, 1999; Anthias and Yuval-Davis 1989; Mani 1990; Sangari and Vaid 1990; Mohanty *et al.* 1991; Chow 1993; Rajan 1993; Barlow 1997).

For much of the post-war period, nationalist issues have been the major political concerns in the immigrant communities in Japan – the Korean community in particular being split between groups with alliance to North or South Korea (Ryang 1998b).[3] Second- and third-generation Korean and Taiwanese residents are now labelled as 'oldcomers' in comparison with the 'newcomers', the newer labour immigrants. Since the 1980s, immigrant communities have protested against the alien registration system, called for full benefits under the welfare system, and protested against their exclusion from public service positions. Some progress has been made on these issues, and the next hurdle is political representation in local government. At the time of writing, some local government areas had created special consultative assemblies for local residents, but had yet to extend voting rights in local government elections to non-Japanese residents.

The second- and third-generation residents are the descendants of an earlier wave of labour migration, but South Korea also contributes significant numbers of illegal 'newcomer' immigrants (Moon 1996). These second- and third-generation residents have more chance of entering the Japanese public sphere through their biculturalism – some are bilingual in Japanese and Korean, while Japanese is the first language for others.[4] Such groups also provide a further dimension to discussion of the situation of the newer immigrant groups, who are gradually becoming more visible in workplaces and in local communities, and who are producing children whose needs will have to be addressed by the welfare and education systems.

The issue of enforced military prostitution (the so-called 'comfort women' issue), which surfaced in the 1990s, removed the possibility of distance from the events of the Second World War and the colonial period, and mobilised Korean resident women in new ways (Yun 1992; Kim 1996; Mackie 2000). The women claiming compensation were often elderly Korean residents surviving in contemporary Japan and the younger women supporting their claim came from the Korean resident community.

In so far as they dealt with relationships of inequality determined by class, gender, ethnicity and 'race', the historical perspectives gained from the campaigns around these seemingly disparate issues have informed recent attempts to understand the situation of immigrant workers in contemporary Japan. The discussion of the situation of immigrant workers in contemporary Japan also draws on the 1970s campaigns around the issue of the links between tourism and prostitution, which revealed the cross-cutting axes of gendered, classed and ethnicised inequalities in the East Asian region. These campaigns drew together networks of activists in Japan, Korea, the Philippines and Thailand (cf. Yamazaki 1996: 95–7).

## Spaces of difference

One of the earliest groups to attempt the ambitious project of understanding these multiple axes of discrimination was the *Shinryaku=Sabetsu to Tatakau Ajia Fujin Kaigi* (Conference of Asian Women Fighting Against Discrimination=Invasion, hereafter abbreviated as Asian Women's Conference), launched in 1970. A major spokesperson of the group, Iijima Aiko, has reflected that the women of the New Left had shifted from working under the guidance of their male comrades to becoming the subjects of their own struggle. These women considered the connections between sexual discrimination and other forms of discrimination, and came to see themselves as both oppressors and oppressed (Mizoguchi *et al.* 1992: 19). While women had been politically active for much of the post-war period, they had rarely been active *as women*, and had failed to challenge or transform the existing political system (Iijima, cited in Mizoguchi *et al.* 1992: 37). The name of their group signalled a consciousness that women in Japan had histories which were linked with those of women in other Asian countries. In other words, they did not simply seek liberation within the bounds of the Japanese nation-state, but situated themselves within the broader history of East Asia.

Another group which attempted to bring the lives of Japanese women into the same frame of reference as other Asian women was the Asian Women's Association (*Ajia no Onnatachi no kai*). The members of this Association chose 1 March 1977 to launch the organisation.[5] They issued a manifesto which outlined their concerns and described their understanding of their relationship with women in other Asian countries:

> We want to express our sincere apologies to our Asian sisters. We want to learn from and join in their struggles.
>
> We declare the establishment of a new women's movement on March first. This day when Korean women risked their lives for national independence from Japanese colonial rule marks the start of our determined efforts.
>
> (*Asian Women's Liberation* 1977: 2)

The first issue of their journal *Ajia to Josei Kaihō* (*Asian Women's Liberation* 1977)[6] focused on political struggles in Korea and included the first of many articles on Japanese women's participation in the colonial project. Subsequent editions focused on issues of political repression which affected women in Asian countries; Japanese cultural imperialism; liberation movements in Asian countries; the economic activities of Japanese companies in Asian countries; working conditions in transnational factories, tourist resorts and plantations; labour activism and its suppression; pollution; the international tourism industry and its links with prostitution; and the history of Japanese activities in Asia (*Ajia to Josei Kaihō* 1977–83).

While the women of the Asian Women's Association sought narratives of liberation in the histories of Third World women, their investigation of the role of women in Japanese history prompted an anxiety about co-optation into state-defined goals. 'Asia' functioned as a space of difference in these discussions. 'Asia' was the space where activists confronted the history of Japanese colonialism and contemporary relationships of inequality in the economic sphere. 'Asia' was the space of political repression, but also the space of political resistance. This was the space where activists imagined a form of resistance they could not find in their own history. Eventually, however, these issues became apparent much closer to home.

In the 1970s, the political economy of tourism came into focus in Japan. The austerity of the post-war years was overtaken by the income-doubling plans of the 1960s. Relationships were normalised with South Korea, while Japanese businesses started to move production offshore, in a reaction against increased wages at home, and increased state control over polluting industries in the aftermath of several industrial pollution incidents. As middle-class Japanese became prosperous enough to become tourists, a particular pattern began to emerge. Statistics on tourist travel to certain Asian countries – first South Korea, then the Philippines and Thailand – showed that an overwhelming majority of these travellers were male. It became apparent that the attraction of these places was the prostitution industry. For these tourists, 'Asia' functioned as a space of illicit sexual practices.

Some of the earliest public demonstrations against this form of tourism were by Korean women who demonstrated in 1973 at the international airport in Seoul (Matsui 1995: 317). Soon they were joined by groups of Japanese women who had come through the student leftist movement – such organisations as the Asian Women's Conference, Christian organisations and newer organisations like the Asian Women's Association. Eventually, the issue brought together networks of activists in Japan, South Korea, the Philippines and Thailand (Mackie 1988, 1998b).

In the 1970s these issues were externalised in as much as this part of the prostitution industry was carried out offshore from Japan, but the issue was very close to home for some women in Japan as they tried to come to terms with the sexuality of the men who engaged in such activities overseas, or who profited from their promotion of such tours. Many women internalised a policing role and spoke of their 'responsibility' to curb the behaviour of Japanese men. More recently, such organisations as the Asian Women's Association have paid attention to Japan's role as a major donor of foreign aid in the region and the increased emphasis on support of tourism-related projects (Mackie 1992). In the 1970s, prostitution was discussed in terms of the tourists who travelled to other Asian countries in search of commodified sexual services. From the 1980s, these services were provided within the major Japanese cities by immigrant women from Southeast Asian countries. In the major Japanese cities, certain spaces have become identified with immigrant workers, bearing such labels as 'Little Bangkok', while particular bars, known as 'Filipina pubs' have become the site for the consumption of the spectacle of sexualised and ethnicised difference.[7]

In academic, popular and activist literature in Japan, 'Asia' is often aligned with the 'feminine'. In popular texts, sexual and romantic encounters serve as allegories for the unequal relationships between a masculine Japan and a feminine Southeast Asia (Kondo 1997; Mackie 1998a). The feminisation of Asia is also reflected in activist literature which collapses different forms of marginality. One book about the so-called '*Japayuki-san*', the immigrant workers from Southeast Asia, bears the subtitle *Asia is a Woman (Ajia wa onna da)* (Yamatani 1985). A recent travel book bears the title, *The Temptation of Asia (Ajia no Yûwaku)*, which adds a further element of sexualisation to the practices of gazing on 'Asia'.

The fashion for 'Filipina pubs' means that the metropolitan gaze on exoticised and eroticised spaces may also be enjoyed in the major urban centres of Japan. What I call the 'metropolitan gaze' is a slight variation on the terminology of Orientalism, which I have elsewhere called the 'orientalising gaze' (Mackie 1998a). The 'orientalising gaze', however, is not purely the privilege of the western male observer. Rather, the gaze of powerful observers on those less powerful is structured in specific relationships of domination and subordination. Given that these relationships can no longer be assumed to fit the pattern of powerful 'western' observers gazing on less powerful 'non-western' objects of the gaze, I prefer to modify the terminology of Orientalism to refer to the 'metropolitan' gaze on colonial, tourist or other specific spaces and bodies.

## The everyday practices of difference

The Asian Women's Association developed a strategy for closing the gap between the women in Japan who read their journal and attended their meetings and the other Asian women who were the focus of these writings and activities. The journal ran a series of articles on the theme of 'Asia in Everyday Life' (*Kurashi no naka no Ajia*). By considering the food, cosmetics and manufactured goods produced in Asia, brought into Japan, and consumed by Japanese women and men, they were able to show the links between themselves and women in other Asian countries (*Ajia to Josei Kaihô* 1981, 1987).[8] The members of the Association thus came to see themselves as benefiting from the exploitation of their Asian sisters, and unable to see these women simply as the objects of pity or concern. By publishing in both Japanese and English, the organisation also made links with international networks.

The early discussion of 'Asia' in everyday life focused on the consumption of products from Asian countries and the dissemination of information about the conditions of their production. From the 1980s, encounters with 'Asia' in everyday life could just as easily mean encounters with the immigrant workers who came to staff bars, small-scale industry and construction sites. These concerns were addressed in a message to potential members of the group:

> For women, Asia at first seems like a world far away, but it's really very close. Bananas from the Philippines, prawns from Indonesia and tea from Sri Lanka have become part of our everyday lives. There has also been an increase in the numbers of immigrant women workers, international students, and brides from Asia. Japanese men go to Asia to buy Asian women, while Japanese companies advance into Asia and employ Asian women as cheap labour. We ask you to join our group, study with us, engage in activism with us, and make links with women in Asia in order to change Japanese society and ask questions about our lifestyles. We welcome people who will support our group as volunteers, because we are a group without hierarchy.
>
> (*Ajia to Josei Kaihô* 1992: back cover)

Thus, by the 1980s, an understanding of the interplay between class, gender and 'race' relations in the sex industry could no longer be externalised as the industry was transformed. Entrepreneurs within Japan started to employ immigrant women from the Philippines, Thailand and South Korea in the bars and massage parlours of Japan's major cities (*Ajia to Josei Kaihô* 1983, 1992; *Asian Women's Liberation* 1991).

As most immigrant workers were illegal, they did not even have the institutional recognition of the German 'guest workers', and thus there was initially no conceptual space for them in discussions of citizenship. Those who entered under the legal visa category of 'entertainer' were marginalised because their labour did not fit accepted categories of productive labour.

Some immigrants, however, crossed a conceptual boundary by marrying into Japanese families. From 1975, the typical 'international marriage' (kokusai kekkon) has changed from the pattern of 'non-Japanese husband/Japanese wife' to 'Japanese husband/non-Japanese wife', with the wives of the latter coming from the Philippines, China, Korea, Thailand, Sri Lanka, Brazil and Peru. There is considerable overlap between supposedly distinct analytical categories. Women who immigrate as brides may be seen as a specific form of labour migration, as they perform housework, childcare, and reproductive and sexual labour in the households they marry into. In other cases, a sham marriage may mask exploitation in the sex industry. Similarly, many legal or illegal immigrants may become involved with Japanese partners. Others come to Japan through the mediation of marriage brokers. For a time, local governments were involved in promoting such marriages, but they now tend to delegate this to private brokers (Kasama 1996: 172).

As these women and their children become part of local communities, notions of community and nationality are being challenged at the grass roots, and commentators on gender relations need ever more sophisticated models of the interplay between gender, class, 'race' and ethnicity. Earlier groups of immigrant workers tended to seek the assistance of shelters in escaping from enforced prostitution and finding a way to return to their home countries. As for immigrant women who marry or live with Japanese men, some seek shelter from domestic violence by husbands or partners, while others seek to regularise the nationality status of their children. In order for the child of a Japanese father and non-Japanese mother to receive Japanese nationality, it is necessary for the father to give official acknowledgement (*ninchi*) of the child (Taylor 1992; Higashizawa, 1993: 161–201).[9] If a woman's visa status is 'spouse of a Japanese national' then divorce will mean that she loses her residence status, and she will need to apply to the Department of Immigration for special permission for continued residence in Japan. While some women may be happy to return to their country of origin in such circumstances, the situation is more complex for women with children.[10]

In 1990, the Immigration Control Act was modified to allow second- and third-generation descendants of Japanese emigrants to enter Japan for up to three years, in a long-term resident category with no restriction on engaging in employment (Yamawaki 1996: 18). These descendants of the Japanese emigrants who migrated to the Americas in the late nineteenth and early twentieth centuries may now enter Japan to work legally under this special visa category, and it seems that some women of these communities are working in Japan in nursing or caring occupations, while others work in factories alongside other family members (Kasama 1996; Yamanaka 1996).[11] The Japanese ethnicity of such immigrants was probably expected to cut down on problems of difference, but local communities must in fact deal with contact between groups with different social and cultural expectations, while schools are addressing the necessity of multilingual and multicultural education for the children of these families.

A concern with 'Asia' has thus shifted from an interest in the conditions of women in other Asian countries to an interest in the situation of women and men from Asian countries working in Japan. Although these immigrant workers were initially seen as a transient phenomenon, they have increasingly become part of Japanese communities through practices of intermarriage. This means that local communities must deal with issues of difference at the level of everyday life.

## The embodiment of difference

In discussing what I call the 'embodiment of difference', it would be easy to be misunderstood. I am not, however, primarily referring to the notion that the different national groups are often thought to be distinguishable in terms of physical differences (hair, eye, skin colour) which are given social meaning in particular contexts. In popular culture in Japan, there is indeed a consciousness of the physical characteristics attributed to Europeans, Africans and other groups; those from other Asian countries, such as the Philippines, are at times described in terms of darker skin colour (Inagaki 1996; Russell 1996; Creighton 1997; Mackie 1998a). Rather, I am interested in the embodied practices which reinforce inequalities. These practices reinforce the connections between immigrant workers and physical or sexualised labour. Embodied practices also reinforce gendered and racialised hierarchies.

A focus on embodied practices can help us to better understand some forms of immigrant labour. But first we need to clarify this notion. The insights gained from analysing particular practices in tourist encounters will be adapted to an understanding of some features of immigrant labour in contemporary Japan. Tourism is often understood simply as a matter of gazing on difference, whether this be exoticised landscapes or exoticised and sexualised individuals. One understanding of the relationship between the tourist gaze and the masculine gaze is formulated by Rojek and Urry (1997: 17), when they state that 'Travel and tourism can be thought of as a search for difference. From a male perspective, women are the embodiment of difference.' While this formulation captures the gendered dimensions of the tourist gaze, it leaves the production of ethnicised and racialised identities through tourist encounters untheorised.

Others have tried to bring the body itself into discussions of tourism and travel (Veijola and Jokinen 1994). Eeva Jokinen and Soile Veijola (1997) invert conventional theorisations of the disembodied, disengaged, displaced traveller by focusing on the 'sextourist', the 'prostitute' and the 'babysitter'. By placing these figures at the centre of their analysis, Jokinen and Veijola help us to understand that the focus on the tourist gaze has obscured the discussion of the embodied politics of travel and sexuality.

In tourist encounters, travellers may expect a range of personalised, and often embodied, services: cooking, cleaning, serving of food and drinks, childcare, entertainment. Some tourists also travel in the expectation of sexual encounters – in the form of casual relationships, or in the form of payment for sexual services. Relationships are never, however, simply encounters between two individuals. Each individual brings a history shaped by the dimensions of gender, class, 'race', ethnicity and sexuality in a field of political and economic inequalities. In tourist discourse, then, we need to go beyond the theorisation of the gaze in order to come to terms with the embodied politics of relationships, to focus, as David Kelly (1997: 72) suggests, on the 'grasp' as well as the 'gaze':

> I suggest then that grasps are at least as well worth thinking about as gazes, as we try to write about power. Equally and more viscerally embodied than gazing, grasp is nevertheless neither more nor less intrinsically material. It is, like gaze, yet in a different way, another vehicle for inscription, embodiment and objectification, for realisation of representations in self and world.[12]

Unequal relationships between guests and hosts – between prosperous First World tourists and less prosperous Third World workers – may also be understood in terms of embodied practices. The embodied practices of the tourism industry, the entertainment industry and the prostitution industry reinforce racialised hierarchies.

Laurie Shrage (1994: 150) suggests that 'the ethnic structure of the prostitute work force in our contemporary world may be conditioned by racial fantasies white people have about themselves and others'. I would replace 'white people' with 'First World', and extend the discussion to the ways in which racialised hierarchies are reinforced through the practices of prostitution, whether these practices take place in Bangkok, Manila or Tokyo. The prostitution industry provides the white-collar citizen with a temporary sexual outlet which does not interfere with his ability to commit to his daily work. Indeed, Anne Allison argues that desire is 'produced in forms that co-ordinate with the habits demanded of productive subjects'; these desires 'make the habitual desirable as well as making escape from the habits of labour seem possible through everyday practices of consumptive pleasure' (Allison 1996: xv; see also Allison 1994). We also need to theorise the production of desire in the context of the eroticisation of ethnicised difference. The desires of the salaryman may be even more effectively managed if the exoticised objects of his fantasies are geographically or conceptually displaced (Mackie 1998a).

Images of otherness in popular texts 'help construct and perpetuate an imagined Japanese self-identity' (Creighton 1997: 212). However, because of recent patterns of labour migration and marriage migration, these 'others' are no longer safely displaced, but are within the boundaries of the Japanese nation-state. Whereas tourists formerly travelled to Bangkok, Seoul and Manila to buy the services of the sexualised entertainment industry, bars and brothels in the major Japanese cities are now staffed by women from these countries. It is thus necessary to displace these 'others' through discursive means. Members of subordinate groups may be included to the extent that their labour can be used, but will often be excluded from full participation in the national community, in a process which Ghassan Hage (1998: 134–8) refers to as the 'dialectic of inclusion and exclusion'.

This concept of embodied practices can also provide an insight into the labour undertaken by immigrant workers in Japan. Under immigration policy, there is an implicit privileging of mental labour over physical labour – a mind/body split whereby intellectual, white-collar work is given recognition, but not manual and physical labour, and certainly not sexual labour. While male workers in construction and manufacturing are often discussed in terms of labour policy, the women who come to work in the entertainment industry are discussed in terms of morality and policing (Mackie and Taylor 1994). So far, the Japanese Department of Immigration has failed to permit immigration for the purpose of engaging in what is defined as unskilled labour. This means that it is illegal to import labour for the purpose of domestic work, although anecdotal evidence suggests that some families are finding ways to employ overseas maids. Japan is thus relatively distinctive in not importing large numbers of domestic workers.[13] Rather, the overwhelming majority of women entering the country from Southeast Asia are working in some part of the 'entertainment' industry. Thus, the sexualised image of immigrant women workers in Japan is partially produced by the workings of immigration policy which makes it difficult for immigrant workers to be employed as domestic workers, while the legal category of 'entertainer' provides a mask for a continuum of sexualised activities from singing and dancing, waitressing and hostessing, to prostitution.[14] In other words, immigrant workers offer a range of personalised, and often embodied, services in the bars and brothels of the urban centres of Japan.

Until around 1988, the majority of illegal immigrant workers were women, who formed the vanguard of the present influx of workers from overseas. However, it was only when large numbers of male workers started to enter the country as unskilled workers that this was seen as a labour issue (cf. Shimada 1994). For male illegal immigrants, their engagement in what is seen as productive labour means that there is a space for discussion of their situation, and even space to consider granting them a limited form of legitimacy, perhaps as 'guest workers'.

While economic rationalist arguments may be made for the recognition of immigrant workers in manufacturing and construction, immigrant women workers often remain beyond the pale of discourses of citizenship. These are the women from Thailand and the Philippines who engage in entertainment, waitressing and prostitution. They are subject to voyeuristic attention, with an added element of sexualisation. While male workers' jobs are physically dirty, these women bear the stigma of sexualised labour. Like their male counterparts, they engage in work with long hours and difficult working conditions, but with the added dangers of violence and sexually transmissible diseases.

The dynamics of the employment of these women in the entertainment industry within Japan has much in common with the purchase of the services of prostitutes in the tourism industry of Southeast Asia. Encounters between male customers in the entertainment industry of the major Japanese cities and the women from Southeast Asian countries working in this industry reinforce gendered and racialised hierarchies through embodied practices. The daily repetition of encounters between Japanese white-collar workers and immigrant workers in the entertainment industries reinforces the opposition between Japanese and non-Japanese, mental labour and physical labour, sexualised labour and non-sexualised labour. Hierarchies are also reinforced through the practice of using immigrant labour in jobs which are thought to be 'dirty, difficult and dangerous': these jobs are seen as increasingly undesirable by young Japanese people entering the workforce, but appropriate to be carried out by immigrant workers.[15]

In some countries arguments are being made for the recognition of prostitution as work, and the decriminalisation of prostitution. It is only quite recently that prostitutes' advocacy groups have been created in Japan (Momocca 1999). While the decriminalisation of prostitution has recently been debated by feminists in Japan (Group Sisterhood 1999) this argument is seldom heard from advocates working with immigrant workers, who recognise the coercive conditions many of these women work under (cf. Ehara *et al.* 1996: 43–8). Discussing prostitution in terms of labour performed is shocking to some, but I would argue that such a perspective is necessary in order to come to terms with the specific situation of these immigrant workers. Discussion of prostitution in terms of labour relations does not imply endorsement of such work, but allows a fuller understanding of the dynamics of the gendered, classed, racialised and ethnicised hierarchies which are reinforced through the practices of sexualised labour.

Many immigrant women work in coercive situations where violence is one of the strategies used to exert control. Workers in shelters attempt to assist women escaping from such violence. Support groups have also addressed the needs of immigrant women charged with violent retaliation against the employers who have coerced them into prostitution. Several Thai women have been charged with murder or attempted murder in the early 1990s, suggesting the desperation of women whose only hope of escape from a coercive situation is violence (Matsui 1992, 1999: 15–18; Matsushiro 1995). These cases provide further insight into the complex gendered dynamics of the situation of such workers. It is tempting to speculate that it is those women who (through linguistic and other barriers) are furthest from the reach of the assistance of non-governmental organisations (NGOs) who find that their only escape is through violent retaliation. Also, it is their immediate supervisors (often immigrant women themselves) who have been the target of such attacks, rather than the entrepreneurs who ultimately profit from their sexual labour.

Domestic violence is another embodied practice which reinforces gendered, sexualised and racialised hierarchies. A distinctive feature of the movement for women's refuges in Japan is that much of the activity in this area was stimulated by the plight of immigrant women. While some public emergency accommodation has existed for most of the post-war period, this has generally been tied to particular legislative programmes, such as welfare legislation or anti-prostitution legislation. Immigrant women in particular were not served by public facilities, and it was private NGOs which created the first refuges in the mid-1980s. Although these refuges were generally set up for immigrant women, they soon had to find room for Japanese women with no other escape from domestic violence. The links between the violence suffered by disparate groups of women brought home the necessity for an analysis of gendered patterns of violence, while the specific vulnerabilities of immigrant women focused attention on the dynamics of gender, class, 'race' and ethnicity. Activists dealing with the violence suffered by immigrant workers and by Japanese and non-Japanese wives in domestic situations have thus been sensitised to what has been called the 'gendered division of violence', whereby those in positions of power due to their gender, class or ethnicity perpetrate violence on those in less powerful positions (Peterson and Runyon 1993: 37; Youngs 1999: 109).

Such an understanding of the 'gendered division of violence' may also be linked with the institutionalisation of prostitution in military institutions, whether this be the history of military prostitution in the Second World War or the current issues of sexual violence against women in the communities adjacent to military bases in Japan, South Korea and other parts of Asia. This is another issue which has brought together feminists in Okinawa and other parts of Japan, South Korea, Thailand and the Philippines (Takasato 1996a, 1996b).

Such encounters between activists in different countries have the potential to challenge what Gupta and Ferguson (1992: 7) have called 'the assumed isomorphism of space, place and culture'. Feminists within Japan can no longer assume that 'Japanese women' are the 'natural' object of their concern, or that they are separated from 'other' Asian women by geographical and cultural boundaries.

## Dealing with difference

The discussion of difference in Japan has moved through several stages, from a time when the dominant assumption was that questions of difference were external to Japan, to a recent recognition that problems of cultural contact and ethnicised difference are now being played out through encounters which take place within the geographical and conceptual boundaries of the Japanese nation-state. Those groups which focus on migration issues have learned that the relationships between immigrants and their relatively privileged hosts in Japan have been shaped by a history of imperialism and colonialism and the features of the contemporary political economy of East Asia. Their experience of dealing with immigration has taught them that issues of difference and inequality are now very close to home, but they have taken their struggles back into international circles through such fora as the United Nations Conference on Women in Beijing in 1995.

The Asian Women's Association has undergone several transformations in the years since its founding, but continues to focus on women and men in Japan and their relationships with women and men in other Asian countries. Their journal is now called *Onnatachi no Nijū Isseiki* (*Women's Asia: 21*).[16] Recent editions focus on the feminisation of poverty, sexual rights, ecology and feminism, and unwaged work, and include translated reports from activists in Thailand and the Philippines. While much of the earlier journal, which appeared under the title

*Ajia to Josei Kaihô/Asian Women's Liberation*, focused on issues of political economy, current editions also include discussion of masculine sexuality and the analysis of cultural representations, once again focusing attention on the politics of everyday life (*Onnatachi no Nijû Isseiki* 1996–2001).

Mainstream media treatment of the issue of immigrant labour is often voyeuristic, describing living conditions, wages and working conditions in fine detail, drawing attention to the co-existence of disparate groups of people in local communities. Other strands of reporting concentrate on foreign residents and their collisions with the criminal justice system, or link immigrant women with the transmission of disease. Such articles draw attention to the co-existence of disparate groups of people within the boundaries of the Japanese nation-state; highlight the exploitation of illegal immigrant workers; and stimulate public discussion on ways of dealing with difference within Japan. At the same time, however, such articles tend to reinforce negative views of difference through a focus on criminality, danger and disease.

Women's groups have been critical of popular men's magazines which publish 'guides' to the brothel areas of Bangkok, Manila and Seoul, Tokyo and regional centres.[17] One strand of this reporting attributes a certain amount of negative agency to the women in the sex industry, describing them as calculating and manipulative; this distracts attention from the structure of the industry and the groups and individuals who profit from their labour. Even articles in relatively well-meaning publications may focus on anecdotes of remittances sent home and electrical goods purchased, rather than the coercive conditions suffered by many women workers (Hinago 1990; Hisada 1990; Mackie 1998a). In contemporary popular culture, representations of Thai and Filipino women are becoming increasingly apparent, but they are generally portrayed solely in the entertainment industry, constructed as body rather than mind, as sexualised 'others', and aligned with the abject: with sexuality, bodily functions and disease (Pollack 1993; Inagaki 1996: 271–96; Kaneko 1996; Buckley 1997; Mackie 1998a).

Women's organisations and citizens' groups have also been critical of the language used to describe immigrant workers. The popular label for women immigrants from Southeast Asia is *Japa-yuki-san* ('women who come to Japan'), a pun on *Kara-yuki-san*, the women who travelled from Japan to Southeast Asia in the late nineteenth century, often being put to work as prostitutes (Mihalopoulos 1993).[18] Today, activists prefer to reject the sexualised connotations of *Japa-yuki-san* and emphasise the commonalities between various groups of illegal immigrants to Japan. Other commentators refer to *Gaikokujin Rôdôsha* (foreign workers), a label which focuses on 'foreign-ness', while sidestepping the distinctions between legal and illegal visa status (Kajita 1994). Advocacy groups prefer to refer to *Kaigai Dekasegi Rôdôsha* (Overseas Migrant Workers), or *Ajiajin Dekasegi Rôdôsha* (Asian Migrant Workers) (Itô 1992: 293). The Immigration Department categorises people according to their visa status: legal or illegal.[19]

Some activist groups have attempted to challenge these representations of immigrant workers by presenting testimonials from the workers themselves (*Asian Women's Liberation* 1991; Tezuka 1992; Higashizawa 1993; Piquero-Ballescas

1994; Okuda and Tajima 1995). Although many of these works are filtered through the commentary of their Japanese authors or editors, they may provide limited space for the words of immigrant workers. Filipino immigrants, through their knowledge of English, have some access to fora where they can speak back to an English-educated Japanese audience, and they have used these fora to protest against negative representations. This is harder to achieve for immigrants from Thailand and other countries: due to linguistic problems, it is Thai women who are said to be subject to the most extreme exploitation as immigrant workers.

Other groups have been involved in setting up shelters for immigrant workers. These shelters may be run by women's groups, Christian groups, citizens' groups with an interest in human rights, and anti-alcohol/anti-drug groups. Most are funded by subscriptions and donations, but some progressive local government areas provide support for private refuges, and complex networks have been built up between NGOs and local government welfare departments. The issue of immigrant workers has also marked a new stage in Japanese feminists' engagement with women from other Asian countries. These 'others' are within the boundaries of the Japanese nation-state, and likely to need assistance with the legal system, or protection from violent pimps or husbands, a major impetus for the setting up of women's refuges.

The authors of a recent report on shelters in Japan comment that there has been a shift in attitude from 'helping' those in need to working towards 'empowerment' in terms decided by the clients themselves (Yokohama-shi Josei Kyôkai 1995). Another commentator discusses initiatives such as Kawasaki City's childcare classes for immigrant women. These classes provide a focus for networking for these women, with a shift from activities led by local welfare workers to activities initiated by the women themselves (Kasama 1996: 177). A representative of Yokohama's Mizura Space for Women describes the development of a new 'human rights service sector' (Abe 1996: 167). Most of the Japanese activity has been carried out by private volunteer organisations dependent on subscriptions and donations. Some have been funded by private foundations, and a few progressive local governments have provided assistance to NGOs. Some commentators are now calling for increased government assistance for such activities, while recognising the dangers of co-optation and loss of autonomy (Kasama 1996: 182).

Volunteers need specific knowledge and linguistic skills in order to deal with various government departments on issues such as living expenses, travel expenses, pregnancy, birth, treatment of illnesses, legal assistance and infant welfare. Most shelters also undertake community education activities such as lectures, educational slide-shows and producing pamphlets, videos or newsletters (Yokohama-shi Josei Kyôkai 1995: 12–13).

The legal situation of immigrant workers is often ambiguous. Article 14 of the Constitution of 1947 states: 'All of the people are equal under the law and there shall be no discrimination in political, economic or social relations because of race, creed, sex, social status or family origin.' There is no clear legal precedent as to whether laws on pensions or welfare apply to non-Japanese residents. Japan has ratified the International Convention on Economic and Social Rights, and the International Convention on Civil, Political, Cultural Rights. Article 2 of the Local Government Law (*Chihô Jichi Hô*) states that local governments have the responsibility for maintaining the safety, health and welfare of residents (Miyajima and Kajita 1996: 9), and some local governments interpret this as a responsibility to address the needs of both Japanese and non-Japanese residents. (Miyajima and Kajita 1996: 11).[20] Much of the casualised work performed by immigrant workers is outside the purview of the Labour Standards Law (*Rôdô Kijun Hô*) of 1947, the major legal instrument concerned with working conditions. Labour unions have until recently been more concerned with permanent, full-time employees in large-scale organisations, and have been relatively blind to the immigrant workers, because of the casual nature of their work, and because their labour is often in unrecognised categories of physical and sexualised labour. Prostitution has been illegal since the passing of the Prostitution Prevention Law (*Baishun Bôshi Hô*) of 1956 (see Mackie 2000).

The recent international focus on gender and human rights has facilitated the development of spaces for the discussion of immigrant labour in Japan and in the international sphere. The concept of women's rights as human rights (Hilsdon *et al.* 2000) links the issue of military prostitution in the Second World War, the problem of militarised sexual violence in Okinawa, and the conditions of illegal immigrant workers in the metropolitan centres (Watanabe 1994; Matsui 1992; Josei no Jinken Iinkai 1994). The United Nations Women's Conference in Beijing in September 1995 provided groups engaged in dealing with immigrant workers with an opportunity to present their issues to an international audience, and to consider strategies for change (*Agora* 1994–5; Matsui 1996).

The Migrant Women's Research and Action Committee produced a booklet for Beijing which outlined their concerns (Migrant Women Workers' Research and Action Committee 1995). In addition to documenting the conditions of immigrant workers, they outlined a programme of action, and argued for support and help for migrant women through autonomous organisations, links between different organisations concerned with migrant labour, legal reform which would allow the application of laws on political and social rights to migrant women, and acknowledgement of the common goals of labour unions and migrant workers (Kadokawa 1995: 54–5).

Such activists have learnt that international relations are based on embodied politics, and that notions of difference are created through the histories which connect people in different nations and through the everyday practices which reinforce differences. They have thus recognised that activism on issues of difference must be carried out with a consciousness of the histories which produce difference, the spaces where differences are produced, the production of difference in the encounters of everyday life, and the embodied practices which reinforce differences.

## Acknowledgements

Research for this chapter was carried out as part of a larger research project on *Citizenship, Gender and Ethnicity in Modern Japan,* funded by the Australian Research Council Small Grant Scheme and the Toyota Foundation, in collaboration with Veronica Taylor and Tessa Morris-Suzuki. This chapter builds on the paper, 'Feminism and the gendered politics of labour migration to Japan', presented at the *International Conference on Women in the Asia Pacific Region: Persons, Power and Politics* in Singapore in August 1997. I am indebted to the audience at that conference for discussion related to these issues. Thanks also to members of the organisations mentioned in this article, and to those friends who provided materials.

## Notes

1   All *overseas residents* in Japan who are there for more than ninety days must be registered in their local government area and carry at all times an Alien Registration Card (*Gaikokujin Tôroku Sho*) which bears a photograph. The law on alien registration was revised in 1993, so that *permanent residents* are no longer required to be fingerprinted (Hanami 1995: 142); under recent amendments temporary residents are no longer fingerprinted.

2   These patterns fluctuate according to the particular agreements on the issue of visas between countries, with immigrants from Iraq, for example, decreasing when visa agreements between the two countries were modified.

3   During the colonial period, Koreans and Taiwanese were treated as 'subjects' of the Japanese Emperor, but Japanese nationality was revoked in the 1950s after the ratification of the San Francisco Peace Treaty. The Korean War made it difficult for these residents to choose to live in Korea, and even those who stayed in Japan had to choose an allegiance to either North or South Korea for the purposes of documentation of their nationality.

4   For a discussion on issues of language maintenance in the Korean community, see Maher and Kawanishi (1995). For examples of Zainichi intellectual production, see such journals as *Zakkyogaku*, which appeared briefly in the 1980s, and *Harumon Bunka*, which has appeared since the 1990s. For a discussion of resident Korean politics and cultural production, see Field (1993).

5   The date 1 March 1919 marks an attempted Korean uprising against Japanese colonial rule, which included the participation of women students (Kim 1982: 233, 260–4).

6   A literal translation of the Japanese title would be 'Asia and Women's Liberation', but 'Asian Women's Liberation' is the title the group uses in English.

7   See my discussion of the representation of these spaces in popular culture in Mackie (1998a). See Silverberg (1997, 1998) for a discussion of such sexualised and ethnicised spaces in Imperial Japan.

8   Cf. Shiozawa (1983) for an attempt to make such material accessible to young readers.

9   In the 1980s, feminist groups, in particular the Asian Women's Association (with the support of socialist Diet members Doi Takako and Tanaka Sumiko) campaigned for the reform of Japan's discriminatory nationality law which allowed only Japanese men to pass on Japanese nationality to their children. Problems often arose with children born of relationships between Japanese women and US servicemen. Some of such children were stateless because of the inability of their mothers to pass on Japanese nationality and the residence requirements which prevented them from gaining US nationality. While the 1985 amendments to the nationality law have removed this particular set of problems, the Japanese father's acknowledgement (*ninchi*) is still required for a child to obtain Japanese nationality.

10 Piquero-Ballescas (1996: 44) suggests that divorce rates between Japanese men and Philippine women may be as high as 70 per cent.

11 Kasama (1996) and Yamanaka (1996) discuss Brazilian and Peruvian women of Japanese descent who work as what they call 'convalescent attendants'.

12 Kelly focuses on the 'gaze' and 'grasp' of power in colonial situations. I have adapted these concepts to a discussion of the embodied practices of gendered and sexualised labour in the tourism industry and in the entertainment industry.

13 Because diplomatic personnel may employ domestic workers or chauffeurs who speak English, this allows them to employ overseas workers, often from the Philippines. While such workers have a legitimate visa status, they do not come under the purview of the Labour Standards Law. The Labour Standards Law is the legislation which regulates the working conditions of regular workers, but does not apply to domestic workers (Azu 1995: 13).

14 As Gupta and Ferguson (1992: 17) note,

> the area of immigration and immigration law is one where the politics of space and the politics of otherness link up very directly. If … it is acknowledged that cultural difference is produced and maintained in a field of power relations in a world always already spatially interconnected, then the restriction of immigration becomes visible as one of the means through which the disempowered are kept that way.

See also Itô (1992: 296) on similar issues.

15 These features are discussed in more detail in Mackie (in press).

16 A literal translation of the Japanese title would be 'Women's 21st Century', but 'Women's Asia: 21' is the title the group uses in English.

17 More recently, such information is accessible through the Internet, as discussed in unpublished papers by Kathleen Maltzahn and Michelle Webb, former students at the University of Melbourne and Curtin University of Technology respectively. The Internet may be seen as yet another space for the consumption of ethnicised and sexualised difference.

18 The word *Kara-yuki-san* is composed of three elements: *Kara*, an archaic word for China but here referring broadly to any overseas destination; *yuki*, from the verb 'to go'; and *san*, an honorific title. *Japa-yuki-san* replaces China with Japan, and thus refers to immigrants who come to Japan. While *san* is a non-gender-specific title, the word *Japa-yuki-san* generally refers to women, and an examination of the context of its usage reveals that it generally refers to Southeast Asian women in sexualised occupations. One can occasionally see male immigrant workers referred to as *Japa-yuki-kun*, using the male-specific title *kun*.

19 I have translated *dekasegi rôdôsha* as 'migrant worker', but the connotation of the phrase is someone who leaves home to work and save money (either within one country or overseas). 'Remittance workers' would be another possible translation.

20 NGOs have been pressuring local governments to ensure that landlords and commercial businesses do not discriminate against non-Japanese. See http://issho.org/BENCI (accessed 5 February 2002). I am indebted to Itô Ruri for clarification of recent developments.

## References

Abe, H. (1996) 'Mizura: providing service to women in Yokohama', in AMPO: Japan–Asia Quarterly Review (eds) *Voices from the Japanese Women's Movement*, New York: M.E. Sharpe.

*Agora* (1994–6) nos 199–217, Tokyo.

*Ajia to Josei Kaihô* (1977–92) nos 1–21, Tokyo.

Allison, A. (1994) *Nightwork: Sexuality, Pleasure and Masculinity in a Tokyo Hostess Club*, Chicago: Chicago University Press.

——(1996) *Permitted and Prohibited Desires: Mothers, Comics, and Censorship in Japan*, Boulder, Colorado: Westview Press.

Anthias, F. and Yuval-Davis, N. (eds) (1989) *Woman, Nation, State*, London: Macmillan.

*Asian Women's Liberation* (1977–91) nos 1–8, Tokyo.

Athukorala, P. and Manning, C. (1999) *Structural Change and Labour Migration in East Asia: Adjusting to Labour Scarcity*, Melbourne: Oxford University Press.

Azu, M. (1995) 'Current situation and problems of Filipino migrant domestic workers', in Migrant Women Workers' Research and Action Committee (eds) *NGOs' Report on the Situation of Foreign Migrant Women in Japan and Strategies for Improvement*, Tokyo: Forum on Asian Immigrant Workers.

Barlow, T.E. (ed.) (1997) *Formations of Colonial Modernity in East Asia*, Durham: Duke University Press.

Buckley, S. (1997) 'The foreign devil returns: packaging sexual practice and risk in contemporary Japan', in L. Manderson and M. Jolly (eds) *Sites of Desire, Economies of Pleasure: Sexualities in Asia and the Pacific*, Chicago: University of Chicago.

Chow, R. (1993)*Writing Diaspora: Tactics of Intervention in Contemporary Cultural Studies*, Bloomington: Indiana University Press.

Creighton, M. (1997) '*Soto* others and *uchi* others: imaging racial diversity, imagining homogeneous Japan', in M. Weiner (ed.) *Japan's Minorities: The Illusion of Homogeneity*, London: Routledge.

Ehara, Y., Nakajima, M., Matsui, Y. and Yunomae, T. (1996) 'The movement today: difficult but critical issues', in AMPO: Japan–Asia Quarterly Review (eds) *Voices from the Japanese Women's Movement*, New York: M.E. Sharpe.

Field, N. (1993) 'Beyond envy, boredom, and suffering: toward an emancipatory politics for resident Koreans and other Japanese', *positions: east asia cultures critique* 1, 3: 640–70.

Group Sisterhood (1999) 'Prostitution, stigma and the law in Japan: a feminist roundtable discussion', in K. Kempadoo and J. Deozema (eds) *Global Sex Workers: Rights, Resistance and Redefinition*, London: Routledge.

Gupta, A. and Ferguson, J. (1992) 'Beyond "culture": space, identity, and the politics of difference', *Cultural Anthropology* 7, 1: 6–23.

Hage, G. (1998) *White Nation: Fantasies of White Supremacy in a Multicultural Society*, Sydney: Pluto Press.

Hanami, M. (1995) 'Minority dynamics in Japan', in J.C. Maher and G. Macdonald (eds) *Diversity in Japanese Culture and Language*, London: Kegan Paul International.

Higashizawa, Y. (1993) *Nagai Tabi no Omoni*, Tokyo: Kaifû Shobô.

Hilsdon, A.M., Macintyre, M., Mackie, V. and Stivens, M. (eds) (2000) *Human Rights and Gender Politics in the Asia Pacific Region*, London: Routledge.

Hinago, A. (1990) 'Japayuki-san no Gyakushû!', in Takarajima Henshubu (eds) *Bessatsu Takarajima 106: Nihon ga Taminzoku Kokka ni Naru Hi*, Tokyo: JICC Shuppankyoku.

Hisada, M. (1990) 'Watashi, Nippon no Papasan to Kekkon shite Happii ne', in Takarajima Henshûbu (eds) *Bessatsu Takarajima 106: Nihon ga Taminzoku Kokka ni Naru Hi*, Tokyo: JICC Shuppankyoku.

Inagaki, K. (1996) 'Nihonjin no Firipin-Zô: Hisada Megumi "Firipina o Aishita Otokotachi" ni Okeru Firipin to Nihon', in A. Isao (ed.) *Ajia no Kôsaten: Zainichi Gaikokujin to Chiiki Shakai*, Tokyo: Shakai Hyôronsha.

Itô, R. (1992) ' "Japayukisan" Genshô Saikô – 80nendai Nihon e no Ajia Josei Ryûnyû', in T. Iyotani, and T. Kajita, (eds) *Gaikokujin Rôdôsha*, Tokyo: Kôbundô.

Jokinen, E. and Veijola, S. (1997) 'The disoriented tourist: the figuration of the tourist in contemporary cultural critique', in C. Rojek and J. Urry (eds) *Touring Cultures: Transformations of Travel and Theory*, London: Routledge.

Josei no Jinken Iinkai (1994) *Josei no Jinken Ajia Hôtei*, Tokyo: Akashi Shoten.

Jung, Y., Oka, M. and Tasaki, H. (1995) 'Gurobaru Feminizumu no Kanôsei', *Imupakushon* 94: 4–24.

Kadokawa, T. (1995) 'Afterword', in Migrant Women Workers' Research and Action Committee (eds) *NGOs' Report on the Situation of Foreign Migrant Women in Japan and Strategies for Improvement*, Tokyo: Forum on Asian Immigrant Workers.

Kajita, T. (1994) *Gaikokujin Rôdôsha to Nihon*, Tokyo: Nihon Hôsô Shuppan Kyôkai.

Kaneko, A. (1996) 'In search of Ruby Moreno', in AMPO: Japan–Asia Quarterly Review (eds) *Voices from the Japanese Women's Movement*, New York: M.E. Sharpe.

Kanô, M. (1995) *Onnatachi no Jûgo* (revised edition), Tokyo: Imupakuto Shuppankai.

Kasama, C. (1996) 'Tainichi Gaikokujin Josei to "Gender Bias" – Nihonteki Ukeire no Ichi Sokumen to Mondaiten', in T. Miyajima, and T. Kajita, (eds) *Gaikokujin Rôdôsha Kara Shimin e: Chiiki Shakai no Shiten to Kadai kara*, Tokyo, Yûhikaku.

Kelly, J.D. (1997) 'Gaze and grasp: plantations, desires, indentured Indians, and colonial law in Fiji', in L. Manderson and M. Jolly (eds) *Sites of Desire, Economies of Pleasure: Sexualities in Asia and the Pacific*, Chicago: University of Chicago.

Kim, P.J. (1996) 'Looking at sexual slavery from a Zainichi perspective', in AMPO: Japan–Asia Quarterly Review (eds) *Voices from the Japanese Women's Movement*, New York: M.E. Sharpe.

Kim, Y.C. (ed./trans.) (1982) *Women of Korea: A History from Ancient Times to 1945*, Seoul: Ewha Women's University Press.

Kitazawa, Y., Matsui, Y. and Yunomae, T. (1996) 'The women's movement: progress and obstacles', in AMPO: Japan–Asia Quarterly Review (eds) *Voices from the Japanese Women's Movement*, New York: M.E. Sharpe.

Kondo, D. (1997) *About Face: Performing Race in Fashion and Theatre*, London: Routledge.

Mackie, V. (1988) 'Division of labour: multinational sex in Asia', in G. McCormack and Y. Sugimoto (eds) *Modernization and Beyond – the Japanese Trajectory*, Cambridge: Cambridge University Press, pp. 218–32.

——(1992) 'Japan and South East Asia: the international division of labour and leisure', in D. Harrison (ed.) *Tourism and the Less Developed Countries*, London: Belhaven Press.

——(1998a) ' "Japayuki Cinderella girl": containing the immigrant other', *Japanese Studies* 18, 1: 45–63.

——(1998b) 'Dialogue, distance and difference: feminism in contemporary Japan', *Women's Studies International Forum* 21, 6: 599–615.

——(2000) 'Sexual violence, silence and human rights discourse: the emergence of the military prostitution issue', in A.M. Hilsdon, M. Macintyre, V. Mackie and M. Stivens (eds) *Human Rights and Gender Politics in the Asia Pacific Region*, London: Routledge.

——(in press) 'Embodiment, citizenship and social policy in contemporary Japan', in R. Goodman (ed.) *Family and Social Policy in Japan*, Cambridge: Cambridge University Press

Mackie, V. and Taylor, V. (1994) 'Ethnicity on trial: foreign workers in Japan', paper presented at the *Conference on Identities, Ethnicities and Nationalities*, July, La Trobe University, Melbourne.

Maher, J.C. and Kawanishi, Y. (1995) 'Maintaining culture and language: Koreans in Ôsaka', with Y.Y. Yi, 'Ikuno-Ku, Ôsaka: Centre of Hope and Struggle', in J.C. Maher and G. Macdonald (eds) *Diversity in Japanese Culture and Language*, London: Kegan Paul International.

Mani, L. (1990) 'Multiple mediations: feminist scholarship in the age of multinational reception', *Feminist Review* 35: 24–41.

Matsui, Y. (1992) 'Ajia ni Okeru Sei Bôryoku', *Kokusai Josei* 6.

——(1995) 'The plight of Asian migrant women working in Japan's sex industry', in K. Fujimura-Fanselow and A. Kameda (eds) *Japanese Women: New Feminist Perspectives on the Past, Present and Future*, New York: The Feminist Press.

——(1996) 'Economic development and Asian women', in AMPO: Japan–Asia Quarterly Review (eds) *Voices from the Japanese Women's Movement*, New York: M.E. Sharpe.

——(1999) *Women in the New Asia: From Pain to Power*, London: Zed Press.

Matsushiro, T. (1995) 'Problems in legal procedures: the murder trial of trafficked Thai women', in Migrant Women Workers' Research and Action Committee (eds) *NGOs' Report on the Situation of Foreign Migrant Women in Japan and Strategies for Improvement*, Tokyo: Forum on Asian Immigrant Workers.

Migrant Women Workers' Research and Action Committee (eds) (1995) *NGOs' Report on the Situation of Foreign Migrant Women in Japan and Strategies for Improvement*, Tokyo: Forum on Asian Immigrant Workers.

Mihalopoulos, B. (1993) 'The making of prostitutes: the Karayuki-san', *Bulletin of Concerned Asian Scholars* 25, 1: 41–57.

Miyajima, T. and Kajita, T. (eds) (1996) *Gaikokujin Rôdôsha Kara Shimin e: Chiiki Shakai no Shiten to Kadai kara*, Tokyo: Yûhikaku.

Mizoguchi, A., Saeki, Y. and Miki, S. (eds) (1992) *Shiryô Nihon Uuman Ribu Shi*, 3 volumes, Kyôto: Shôkadô.

Mohanty, C.T., Russo, A. and Torres, L. (eds) (1991) *Third World Women and the Politics of Feminism*, Bloomington: Indiana University Press.

Momocca, M. (1999) 'Japanese sex workers: encourage, empower, trust and love yourselves!', in K. Kempadoo and J. Doezema (eds) *Global Sex Workers: Rights, Resistance and Redefinition*, London: Routledge.

Moon, O. (1996) 'Migratory process of Korean women to Japan', in International Peace Research Institute Meiji Gakuin University (eds) *International Female Migration and Japan: Networking, Settlement and Human Rights*, Tokyo: International Peace Research Institute Meiji Gakuin University.

Okuda, M. and Tajima, J. (1995) (eds) *Shinban: Ikebukuro no Ajia Kei Gaikokujin*, Tokyo: Akashi Shoten.

*Onnatachi no Nijû Isseiki* (1996–2001) Tokyo.

Peterson, V.S. and Runyon, A.S. (1993) *Global Gender Issues*, Boulder, Colorado: Westview Press.

Piquero-Ballescas, M.R. (1994) *Firipin Josei Entatênâ no Sekai*, Tokyo: Akashi Shoten (translation of Piquero-Ballescas, M.R. (1993) *Filipino Entertainers in Japan: An Introduction*, Quezon: Foundation for Nationalist Studies).

——(1996) 'The expanding Ds and the Filipino women in Japan', in International Peace Research Institute Meiji Gakuin University (eds) *International Female Migration and Japan: Networking, Settlement and Human Rights*, Tokyo: International Peace Research Institute Meiji Gakuin University.

Pollack, D. (1993) 'The revenge of the illegal Asians: aliens, gangsters, and myth in Ken Satoshi's "World Apartment Horror"', *positions: east Asia cultures critique* 1, 3: 676–714.

Rajan, R.S. (1993) *Real and Imagined Women: Gender, Culture and Postcolonialism*, London: Routledge.

Rojek, C. and Urry, J. (1997) 'Transformations of travel and theory,' in C. Rojek and J. Urry (eds) *Touring Cultures: Transformations of Travel and Theory*, London: Routledge.

Russell, J.G. (1996) 'Race and reflexivity: the Black Other in contemporary Japanese mass culture', in J.W. Treat (ed.) *Contemporary Japan and Popular Culture*, Richmond: Curzon Press.

Ryang, S. (1998a) 'Love and colonialism in Takamure Itsue's feminism: a postcolonial critique', *Feminist Review* 60: 1–32.

——(1998b) 'Nationalist inclusion or emancipatory identity? North Korean women in Japan', *Women's Studies International Forum* 21, 6: 581–97.

Sakamoto, Y. (1996) ' "Jinmin no Umi" no naka ni Dassô Hei tachi wa ita – Beheiren Dassô Hei Enjo Katsudô no koto', *Jûgoshi Nôto* 8: 172–5.

Sangari, K. and Vaid, S. (eds) (1990) *Recasting Women: Essays in Indian Colonial History*, New Brunswick, New Jersey: Rutgers University Press.

Shimada, H. (1994) *Japan's 'Guest Workers': Issues and Public Policies*, Tokyo: University of Tokyo Press.

Shiozawa, M. (1983) *Mêdo in Tônan Ajia: Gendai no Jokô Aishi*, Tokyo: Iwanami Junia Shinsho 60.

Shrage, L. (1994) *Moral Dilemmas of Feminism*, London: Routledge.

Silverberg, M. (1997) 'Remembering Pearl Harbor, forgetting Charlie Chaplin, and the case of the disappearing western woman: a picture story', in T. Barlow (ed.) *Formations of Colonial Modernity in East Asia*, Durham: Duke University Press.

——(1998) 'The café waitress serving modern Japan', in S. Vlastos (ed.) *Mirror of Modernity: Invented Traditions of Modern Japan*, Berkeley: University of California Press.

Singhanetra-Renard, A. (1996) 'Networks for female migration between Japan and Thailand', in International Peace Research Institute Meiji Gakuin University (eds) *International Female Migration and Japan: Networking, Settlement and Human Rights*, Tokyo: International Peace Research Institute Meiji Gakuin University.

Spivak, G.C. (1987) *In Other Worlds: Essays in Cultural Politics*, New York: Routledge.

——(1990) *The Post-Colonial Critic: Interviews, Strategies, Dialogues*, New York: Routledge.

——(1993) *Outside in the Teaching Machine*, New York: Routledge.

——(1999) *A Critique of Postcolonial Reason*, Cambridge, MA: Harvard University Press.

Stalker, P. (1994) *The Work of Strangers: A Survey of International Labour Migration*, Geneva: International Labour Office.

Takasato, S. (1996a) 'The past and future of Unai, sisters in Okinawa', AMPO: Japan–Asia Quarterly Review (eds) *Voices from the Japanese Women's Movement*, New York: M.E. Sharpe.

——(1996b) *Okinawa no Onnatachi: Josei no Jinken to Kichi. Guntai*, Tokyo: Akashi Shoten.

Tanaka, K. (1995) 'The new feminist movement in Japan, 1970–1990', in K. Fujimura-Fanselow and A. Kameda (eds) *Japanese Women: New Feminist Perspectives on the Past, Present and Future*, New York: The Feminist Press.

Taylor, V. (1992) 'Law and society in Japan: does gender matter', in V. Mackie (ed.) *Gendering Japanese Studies*, Melbourne: Japanese Studies Centre.

Tezuka, C. (1992) *Tai kara kita Onnatachi: Sabetsu no Naka no Ajia Josei*, Tokyo: San'Ichi Shobô.

Veijola, S. and Jokinen, E. (1994) 'The body in tourism', *Theory, Culture and Society* 11: 125–51.

Watanabe, K. (ed.) (1994) *Josei, Bôryoku, Jinken*, Tokyo: Gakuyô Shobô.

Yamanaka, K. (1996) 'Factory workers and convalescent attendants: Japanese–Brazilian women and their families in Japan', in International Peace Research Institute Meiji Gakuin University (eds) *International Female Migration and Japan: Networking, Settlement and Human Rights*, Tokyo: International Peace Research Institute Meiji Gakuin University.

Yamatani, T. (1985) *Japayuki-san: Ajia wa Onna da*, Tokyo: Jôhô Sentâ Shuppankyoku.

Yamawaki, K. (1996) 'An overview of the influx of foreign workers to Japan', in International Peace Research Institute Meiji Gakuin University (eds) *International Female Migration and Japan: Networking, Settlement and Human Rights*, Tokyo: International Peace Research Institute Meiji Gakuin University.

Yamazaki, H. (1996) 'Military slavery and the women's movement', in AMPO: Japan–Asia Quarterly Review (eds) *Voices from the Japanese Women's Movement*, New York: M.E. Sharpe.

Yokohama-shi Josei Kyôkai (eds) (1995) *Minkan Josei Sherutaa Chôsa Hôkoku 1: Nihon Kokunai Chôsa Hen*, Yokohama: Yokohama Shi Josei Kyôkai.

Youngs, G. (1999) *International Relations in a Global Age: A Conceptual Challenge*, Cambridge: Polity Press.

Yun, J.O. (ed.) (1992) *Chôsenjin Josei ga Mita 'Ianfu Mondai'*, Tokyo: San'Ichi Shobô.

# 11 Sites of transnational activism

## Filipino non-government organisations in Hong Kong

*Lisa Law*

## Introduction

Theorising transnational labour migration in a 'global', 'post-colonial' and 'late capitalist' world is a complex undertaking. Somehow less decipherable in terms of clear colonial or imperial histories, or through a hegemonic frame of global capitalist development, it is difficult to make presumptions about what propels millions of people to transgress national borders in search of greener pastures. The conditions of much of this migration are, of course, economic. The 'miracle' economies of East and Southeast Asia have generated much intra-regional migration throughout the 1980s and 1990s, for example, and the Asian financial crisis of the late 1990s has prompted several governments to alter their policies on guest workers. Yet a purely economic analysis of labour migration tends to eschew more nuanced understandings of the intersections between cultural and economic politics in the Asia-Pacific region. Given that labour migration has also seen the increased importance of hundreds – if not thousands – of political organisations, there is a need to ask different questions about how labour migration might also be transforming Asian cultural politics. These organisations assert pressure for social change which transgress clear national boundaries, and the networks produced through advocacy programmes provide interesting insights to emerging political spaces where activists negotiate the meanings of transnational labour migration.

Drawing on recent writings about international politics and diasporic culture, this chapter examines the activities of non-government organisations (NGOs) which do advocacy work on behalf of Filipino women who migrate to Hong Kong to seek employment as domestic workers. Exploring NGO images of these labour migrants in Asia, and how they are conceptualised and circulated, I examine emerging spaces of transnational cultural production in the realm of political activism. Filipino NGOs have advocated changes to government policies that facilitate the migration of Filipino women since the institution of the government's Overseas Employment Program in 1972.[1] In the 1990s and beyond, Filipino labour migrants are found around the world, NGOs work at local, bi-national and global scales, and their activities stretch far beyond lobbying the Filipino government or politicising simple push–pull factors of migration. Furthermore, the financial crisis in Asia has prompted the Hong Kong government to decrease the minimum wage for domestic workers, which in turn has facilitated new 'anti-wage cut' coalitions between Filipino NGOs and Thai, Sri Lankan, Nepalese and Indonesian organisations. Indeed, NGO research and advocacy draw together a diverse range of feminist and nationalist convictions, and constitute more of a transnationally inflected movement than a hierarchical strategy of activism. Here I elaborate NGO networks and activities within a framework of cultural politics.

The chapter is organised in three parts. First, I give a brief background to labour migration from the Philippines, describing some of the government policies that have facilitated overseas contract work since the early 1970s. I emphasise how historical, institutional and economic factors have combined to make overseas employment an opportunity for ordinary Filipinos, and how labour migration has taken on specific trajectories for Filipino women. Second, I examine NGO activist discourse that emerged as a resistance to the government's strategy of exporting Filipino labour, and describe three political campaigns launched by Filipino NGOs in Hong Kong. While in some respects these campaigns remain bound to the discursive sphere of national politics, they also transcend the territorial boundaries that delineate flows of labour migration. What I aim to demonstrate in the third part of the chapter is the extent to which these discussions constitute a post-national/diasporic public sphere, while at the same time articulating the importance of nation in the cultural politics of labour migration.

In order to capture the transforming terrain in which Filipino NGOs conduct their activities, it is necessary to recast labour migration in terms which acknowledge a world of global flows at the end of the twentieth century. Appadurai's (1996) conception of 'ethnoscapes' is helpful in contextualising NGO activism within a broader frame of diasporic politics that account for the flow of ideas as well as people. Appadurai (1996: 33) suggests that ethnoscapes are 'landscapes of persons who constitute the shifting world in which we live: tourists, immigrants, refugees, exiles, guest workers and other moving groups and individuals'. At the same time, they are also 'deeply perspectival constructs' which mobile populations create to make sense of such deterritorialised worlds (Appadurai 1996: 33). Without the nation as a primary point of identification, the subjectivity of deterritorialised groups brings together diverse sets of people, ideas and cultures. As a result, ethnoscapes are 'no longer familiar anthropological objects, insofar as groups are no longer tightly territorialised, spatially bounded, historically unselfconscious, or culturally homogeneous' (Appadurai 1996: 48).

Understanding Filipino NGOs within this context enables a new problematic for approaching their activities as they interact with different people, nations and cultures in Hong Kong. It also moves away from hierarchical conceptions of scale – conceptions that hierarchically locate the local and the global. By focusing instead on ethnoscapes of activism, it is possible to move away from conceptions of 'local politics' which do not account for the multiple and shifting scales of politics within which most organisations work. Hong Kong is only one site in a landscape of labour migration, and the perspectives of NGOs provide a productive example of the tensions between nationalism and transnationalism in the global cultural economy.

# Economic heroes of the Philippines

The deployment of overseas contract workers (OCWs) is an important but contested strategy in contemporary economic development policies in the Philippines. In 1997 there were somewhere between 2 and 6.6 million Filipinos working out of the country, and annual remittances from OCWs were estimated at US$4.3 billion in 1996 (May and Mathews 1997: iii). This is a major source of foreign currency for the national economy and helps to raise the standard of living for thousands of Filipino families. While the financial gains of a labour-exporting economy are clear, the negative consequences of labour migration are palpable. Many migrant workers suffer ethnic and gender discrimination in the countries where they work (for example, poor working conditions, abusive employers), and in the Philippines there is evidence of domestic stress (for example, family and marriage breakdown).[2]

Labour migration from the Philippines is a phenomenon of the twentieth century, and is generally understood as occurring in three separate but inter-related waves (Tolentino 1996; Gonzalez 1998). The first wave took place at the turn of the last century – shortly after the United States assumed colonial power in the Philippines – when many Filipinos migrated to Hawaii to fill a demand for plantation labour. Later, migrant labourers travelled to California on the main-land to fill the growing need for orchard workers. In both Hawaii and California, Filipinos formed tightly knit communities that comprise an important part of American multicultural society today. The second wave of labour migration began after the Second World War and continued into the 1970s, and is commonly known as the 'brain drain' from the Philippines. Many professionals migrated to the United States, to sites of American military bases (for example, Guam, Okinawa and the Wake Islands) and to countries such as Canada and Australia. Here they took up permanent residence, while at the same time main-taining strong cultural links to the Philippines. The importance of this group to the Philippine economy was acknowledged in a *balikbayan* (or 'returning national') programme which offered discounted airfares and tax incentives to encourage overseas Filipinos to visit home, spend money as tourists and invest in the Philippine economy. This programme had the effect of transforming over-seas Filipinos into tourists in their own country and, as Rafael (1997: 273) suggests, *balikbayans* came to represent 'the nation's failure to materialise itself as the locus of people's desire' (see also Aguilar 1996).

The third wave of labour migration came with the Overseas Employment Program of 1972, and took on the new form of 'overseas contract work'. Filipino men and women no longer migrated to work with the possibility of future settlement, and were more likely to engage in construction or domestic work rather than medical or other professions. This third wave began in the 1970s, when the oil boom in the Middle East created new employment opportu-nities for both skilled and semi-skilled labour, and by the early 1980s the 'success' of the programme fostered systematic policy development for overseas employ-ment. In 1982, the government established the Philippine Overseas Employment Administration (POEA), which was conveniently timed to take advantage of the economic boom in East and Southeast Asia. Employment opportunities became available in countries such as Japan, Taiwan, Malaysia, Singapore and Hong Kong, and this trend has continued into the 1990s and beyond. While the second wave of migration from the Philippines resulted in a 'brain drain' from the country, this third wave has created the problem of 'de-skilling'. Not only are skilled Filipinos taking up employment overseas – therefore creating a deficit of skilled workers at home – but many professionals (for example, teachers) have given up their professions in the Philippines in order to earn larger salaries from less skilled jobs abroad.

While *balikbayans* and overseas contract workers share the experience of being overseas Filipinos, their status outside as well as within the Philippines is qualitatively and institutionally different. *Balikbayans* tend to occupy identity positions such as 'hyphenated Americans' in multicultural societies and are encouraged to visit and invest in the Philippines (Rafael 1997). Overseas contract workers, on the other hand, are temporary residents in countries which may not have institutional frameworks to protect temporary foreign nationals, and are understood as bolstering the Philippine economy through their regular remittances home. The importance of remittances was made clear in 1982 when the government instituted Executive Order 857, which forced overseas workers to remit 50 per cent of their earnings through Filipino banks. While the Order proved too difficult to enforce, it does reveal the reliance of the Filipino government on incomes earned overseas. Furthermore, it was in 1988 that President Cory Aquino, speaking to a group of domestic workers in Hong Kong, first coined the term 'national heroes' for overseas workers, a term which is mobilised in national rhetoric on a regular basis (Rafael 1997: 274). As Rafael (1997: 276) suggests:

> By encoding OCWs as national heroes, Aquino and her successor, Fidel Ramos, have sought to contain the anxieties attendant upon the flow of migrant labour, including the emotional distress over the separation of families and the everyday exploitation of migrants by labour contractors, travel agents and foreign employers.

This latest wave of migration from the Philippines has created different employment opportunities for men and for women. While men have found employment in the construction or seafaring industries, women have tended to work in the entertainment or domestic service industries. Very little has been written about the experience of male overseas workers (although see Margold 1995), but there is a burgeoning literature which documents the phenomenon of overseas Filipino women, particularly in the Asia-Pacific. Many researchers have elaborated the experiences of domestic workers in Hong Kong, Singapore and Vancouver (Huang and Yeoh 1996; Constable 1997; Pratt 1997), entertainers in Japan (Mackie 1994, 1998; Tyner 1996; Buckley 1997), and mail-order brides in Australia (Holt 1996; Robinson 1996). There is also evidence that these women are frequently stereotyped as prostitutes wherever they are, indicating the highly sexualised perception of Filipino migrants (Lowe 1997). These perspectives highlight the importance of the receiving country's attitudes towards gender, ethnicity and culture, as well as the national policies towards foreign domestic workers, in both the economic and the social lives of Filipino women.

Using the specific example of Filipino women who migrate to Hong Kong, it is possible to see how history, economics and institutional policy combined to encourage more than 100,000 Filipino women to engage in domestic work in the mid- 1990s. Migration to Hong Kong began in the 1970s, when there was a convergence between the Philippines Overseas Employment Program and a shortage of Chinese domestic workers in Hong Kong. Prior to the turn of the last century, an overwhelmingly male population dominated Hong Kong, and the so-called servants of this time were men. In the early 1900s, and with the increasing numbers of Chinese and British women and children migrating to the colony, the demand for female household workers increased. Men moved into coolie labour and women took their place as domestics. The women employed as servants were initially *muijai*, young girls of around 10 years old, but this practice ended in the 1920s. In the 1940s, there was an influx of women migrating from mainland China, and after the Second World War there was an influx of refugees. Until the 1970s, these latter two groups – the famed Chinese *amahs* – supplied the labour market for domestic help (for a review of this history, see Constable 1997). The domestic service industry had therefore been dominated by live-in *amahs* for most of the century, but in the 1970s these women were attracted to the numerous, lucrative and more autonomous service and industry jobs which became available as Hong Kong joined the ranks of Asia's 'miracle' economies.

It was also at this time that Filipino women were being encouraged to work abroad, and provided a practical alternative to Chinese *amahs*. Partly due to its geographic proximity, but also due to an instituted minimum wage, Hong Kong has been a favoured destination for Filipino women seeking overseas employment. In 1998, domestic workers earned the minimum wage of HK$3,860 (US$497),[3] making this type of employment an attractive option even for women with college and university degrees. Between 1975 and 1991, the number of Filipino domestic workers rose from 1,000 to 66,000, and stood at 120,000 in the mid-1990s. Filipinos currently form the largest non-Chinese community in Hong Kong. Indeed, Filipinos have dominated the foreign domestic labour market in Hong Kong since the 1970s, although there are an increasing number of Indonesian, Thai and Sri Lankan women.[4]

## Fields of activist discourse

> If the OCWs are the unsung heroes of Philippine economic development, NGOs are the unsung heroes of OCW welfare and protection.
>
> (Gonzalez 1998: 139)

The maltreatment of overseas Filipino women gained popular press in 1995 when Flor Contemplacion, a domestic worker in Singapore, was convicted of murder and sentenced to death. Flor had been accused of murdering Delia Maga, another Filipino domestic worker on contract in Singapore, and the child of Maga's Singaporean employer. Although Flor confessed to the murders in 1991, she later claimed to have been forced to confess under duress. Further evidence suggested that the causes of Maga's death might have been questionable. Despite disagreements over the evidence, it was widely believed that Flor had been wrongly accused, and the Filipino nation mobilised around the issue, demanding detailed reports about the handling of the case, the severing of relations with Singapore and the boycott of Singaporean goods (May 1997; Gonzalez 1998). Flor's execution and funeral were broadcast on live television, and candle-lit vigils were held around the country, as well as in cities such as Hong Kong and Vancouver.

Flor's case not only outraged the nation, it became the focus of discussion for feminist–nationalist and migrant organisations based in the Philippines and abroad (for example, General Assembly Binding Women for Reform, Integrity, Equality, Leadership and Action (GABRIELA) and Migrante International).[5] While some NGOs suggested that the hurried resolution of the Contemplacion case demonstrated the government's desire to normalise economic ties with Singapore, others emphasised how the case was able to bring issues of labour migration into national and international debates (Gonzalez 1998: 9). Activists used Flor's case to highlight the abuse of Filipino women abroad, the complicity of economic policies in the abuse of overseas workers, and the inability of Philippine embassies and consulates to protect or monitor the rights of Filipino nationals abroad. Flor was, after all, only one of hundreds of Filipinos on death row in foreign countries, and Filipino families had been 'witnessing OCW casket after casket and hearing one OCW horror story after another' for almost two decades (Gonzalez 1998: 6). In general, NGOs emphasised the adverse effects of an outward-looking economy on women, a perspective that has been prevalent since the implementation of a formal overseas contract programme by President Marcos in the 1970s.

Although Flor's story popularised the plight of domestic workers in the national and international media, women working in Hong Kong have long been the subject of much research, advocacy work and debate. Several NGOs in Hong Kong have offices and drop-in centres that have monitored the working and living conditions of Filipino women since the 1970s. These organisations have taken up issues such as illegal recruitment and the exploitation of women through placement agencies, the underpayment of salaries, minimum wage rises/cuts and other government policies affecting domestic workers, as well as more serious cases of physical and psychological abuse. NGOs also spend large proportions of their time providing paralegal assistance to domestic workers involved in legal disputes with their employers. It was not until the Contemplacion case that the government was forced to take these issues seriously, and the Migrant Workers and Overseas Filipinos Act of 1995 strengthened the role of the Overseas Workers Welfare Association (OWWA) to provide more services for overseas workers as well as creating migrant resource centres in Philippine consulates and embassies (May and Mathews 1997; Gonzalez 1998). While OWWA offices have responded to the demands of NGOs by providing services such as shelters for migrant workers in crisis, NGOs remain the primary focus of welfare provision in Hong Kong (in contrast to Singapore, where OWWA and depoliticised religious organisations fill the gap – see Yeoh and Huang 1999). This is partly due to the sheer numbers of women requiring assistance, and partly because NGOs have more flexible contact hours with less bureaucratic paperwork. It is also because women's perceptions of OWWA are enmeshed with government fees collected from overseas workers, and because NGOs are better placed to publicise issues which are sensitive to both Philippine and Hong Kong governments.

The importance of overseas contract work to the Philippine economy, and the adverse conditions these women face in terms of working conditions, living quarters, abusive employers and so on, have all been well documented (Paz Cruz and Paganoni 1989; David 1991; Lane 1992; de Guzman 1994; Tyner 1996). The role of NGOs in preparing migrants for overseas migration, helping migrants with their daily problems, and influencing policies on overseas labour have also been described (see Santos 1989; Battistella 1993; Diaz 1993). Of concern in this chapter, however, is how the activities of Filipino NGOs are reflective of broader changes in the constitution of national and transnational politics. To begin with, NGOs comprehend the Philippines as an unequal player in the global economy. Repaying loans to the International Monetary Fund and World Bank is an urgent priority for contemporary governments, which must find sources of foreign exchange to repay these debts. One strategy is to obtain foreign currency through overseas workers' remittances, but in so doing the Philippine government commodifies its workforce. Leftist women's groups claim overseas women are exploited victims in a global economy dominated by these foreign agendas, reproducing prior colonial and neo-colonial relations. Furthermore, as the economies of Asia undergo transformation, the feminisation of labour migration is understood as particularly disadvantaging to women from the so-called 'developing countries'. Well-educated Filipino women are forced to accept semi- and unskilled jobs abroad, for example, because social mobility is premised on their circulation in international labour markets. In other words, Filipino women escape the 'dysfunctional' national space to circulate in a transnational space where they suffer a triple bind: 'first as a foreigner, second as a woman in patriarchal societies, [and] third as a woman working in professions regarded as menial and even socially undesirable' (Tolentino 1996: 58).

The perspectives of NGOs are important in terms of popularising the plight of overseas women and, perhaps more importantly here, articulating geopolitical relations in the Asia-Pacific region. As a major labour-sending country in the region, the Philippines participates in the growing economies of Asia without becoming a 'miracle' economy itself. Filipino women tend to fill the gap where local populations are unable or simply unwilling: in domestic service and entertainment industries. A feminised labour trade also means that all Filipino women become equated with 'maids' or 'entertainers' (often a euphemism for prostitute), and the nation itself assumes a regional image as a supplier of contract labour. The history of the NGO movement in Hong Kong, which I detail below, provides insights into how NGOs have localised these perspectives and responded to policies and programmes which adversely affect Filipino women.

As mentioned above, in 1982 the Filipino government instituted Executive Order 857 which forced overseas contract workers to remit 50 per cent of their earnings through Philippine banks. Domestic worker organisations in Hong Kong were infuriated by this infringement on the rights of overseas workers, and by 1984 had formed an alliance of ten organisations called United Filipinos Against Forced Remittance (Constable 1997). The alliance was instrumental in having the Executive Order revoked and, with such success to their credit, the coalition renamed itself United Filipinos in Hong Kong (UNIFIL). UNIFIL remains an umbrella organisation for approximately twenty-five NGOs that monitor the working and living conditions of domestic workers in Hong Kong, and has spearheaded several campaigns addressing the protection of migrant workers' rights. Three of these campaigns address (1) Hong Kong immigration policies; (2) fees imposed by the Philippine government; and (3) changes to the instituted minimum wage for domestic workers. Each campaign demonstrates different strategies of NGO activism which connect labour migration to the government policies of the Philippines and Hong Kong, as well as to broader issues of migrants' rights.

Under the New Conditions of Stay, a Hong Kong immigration policy instituted in 1987 and commonly known as the 'Two Week Rule', foreign domestic workers who break their contracts before the end of a two-year period must return to their country of origin within a period of two weeks. This policy was instituted to prevent foreign workers from job-hopping, or from remaining in Hong Kong to work illegally. Given the expenses – including passport, visa, travel and agency fees – incurred to travel abroad, however, many workers find these conditions stressful, particularly if their contract is broken because of abusive employers, or by employers who simply find the domestic helper unsuitable for their family's needs. Only in exceptional circumstances will the Immigration Department allow domestic workers to remain in Hong Kong to seek another employer, since the right to stay is entirely dependent upon the status of employment contracts (Tellez 1993). NGOs took up this campaign to publicise the adverse effects of this policy, particularly in terms of domestic workers who stay with their employer despite poor working conditions until their contracts are complete. In a primer to the campaign, the Mission for Filipino Migrant Workers (MFMW) framed the issue in the following way:

> According to the International Bill of Human Rights of 1948, all human beings are guaranteed the right to free choice of work, and to protection against unemployment (Article 23). And when their human rights are

violated, all persons have the right to an effective remedy by the competent national tribunals (Article 8). According to the new 1990s charter of the International Convention on the Protection of All Migrant Workers and Members of their Family, 'migrant workers shall enjoy equality of treatment with nationals in respect of protection against dismissal, unemployment benefits, access to public work schemes ... and access to alternative employment' (Article 54). Yet despite these guarantees, and their reiteration in the Hong Kong Bill of Rights, domestic workers in Hong Kong continue to face abuses of various kinds, or to face legal recourse which can take weeks, months, or even a year, and which may result in their financial ruin.

(MFMW 1996: n.p.)

There are a number of problems with the New Conditions of Stay that NGOs such as the MFMW have documented through aiding domestic workers in times of crisis. First and foremost is the problem of domestic workers who stay with abusive employers in order to avoid being repatriated. While there are exceptions to the New Conditions of Stay, domestic workers who choose to report abusive situations to the Labour and Immigration Department often face a year of hearings and tribunals during which time they are not permitted to work. Furthermore, there are no obligations for the employer of a domestic worker to provide the airfare back to the Philippines (which is a condition of the end of contract) or to pay any remaining salary due within two weeks. In short, NGOs claim these conditions violate the international conventions and Bills of Rights which should protect domestic workers, and highlight how unequal relations between domestic workers and their employers are enshrined in government policy.

Women's human rights were high on the agenda at the 1995 Fourth World Conference on Women in Beijing. In parallel NGO meetings, organisations such as MFMW provided case studies of Filipino women in Hong Kong describing how processes of economic globalisation were facilitating labour migration in general and discrimination and abuse in particular. In the Statement of the NGO Migrant Caucus, NGOs urged all countries to commit to United Nations conventions on the protection of migrant workers' rights and – in a rather specific credo for a general document – stressed the importance of granting domestic workers a legal status independent of their employers since employment contracts had the effect of creating abusive dependencies for overseas workers (Asian Migrant Centre (AMC) 1995: 45). NGOs such as the MFMW integrated the specificities of Hong Kong immigration policy with global discussions about human rights, encouraging conference participants to view foreign labour policies as an important women's issue. In this way, the MFMW's documentation was simultaneously aimed at changing the New Conditions of Stay as well as integrating migrant workers into the Platform for Action in Beijing.

The second campaign relates to the fees that the Philippine government collects from overseas contract workers. In Hong Kong, there has been an ongoing campaign to challenge all new fees introduced by the Philippine government, but in the latter half of 1998, an alliance of eighty-three migrant worker organisations joined together to form the Coalition Against Government Exactions (CAGE). CAGE was formed to address the imposition of excessive fees through the Philippine Consulate, and initiated a campaign to reduce the fee required for authenticating employment contracts. Their efforts saw the Authentication Fee reduced from HK$425 to HK$255. Most recently, attention in Hong Kong has been directed at the Philippine government's Memorandum of Understanding No. 8 (MOI#8). Under MOI#8, a suggested fee of HK$200 will be required from overseas workers in order to process new contracts at the job site (i.e. in Hong Kong). MOI#8 proposes that this fee be collected by the OWWA office, and be used to upgrade its welfare services. UNIFIL was successful in stopping the implementation of MOI#8 in 1996, particularly since the OWWA shelter in Hong Kong was proven to be grossly underfunded: it was severely overcrowded and residents were not provided with adequate daily sustenance. As a member of CAGE, UNIFIL has continued its campaign against the government of President Estrada, arguing that these policies continue to commodify Filipino workers (UNIFIL 1999). The fees collected under MOI#8 are in addition to fees already paid, such as visa and passport renewals, recruitment agency fees, Authentification Fees and Overseas Employment Certificates, as well as the Mandatory Insurance and Repatriation Bond and Mandatory Medicare. UNIFIL–Migrante, a collection of NGOs based in the Philippines, Hong Kong and around the world, issued a position paper on fees that stated:

> History and practice have shown how the Philippine government through its agencies like the POEA and OWWA have further exploited the economic capacities of Filipino OCWs. Time and again they have invented, imposed and collected fees – for repatriation, insurance, medical care, etc. Yet in reality these fees are not used for their intended purposes ... OWWA should have, by now, US$162.5 million in its Welfare Fund ... with such a huge amount, OWWA could have delivered the best service and looked into the welfare of Filipino OCWs working in some 168 countries around the world. How can you call it quality service when OWWA has only sixteen centres around the world?
>
> (UNIFIL–Migrante 1996: n.p.)

This position paper highlights the complicity of the Philippine government in commodifying its workforce and, rather than effecting measures to protect migrant workers, being more concerned with financial gain. Furthermore, UNIFIL (1999) suggests that the government is placing Filipino women in 'double jeopardy' by the imposition of new fees, since the institution of MOI#8 comes at a time when the Hong Kong government has lowered the minimum wage (discussed below). Drawing on the experiences of NGOs working for migrants around the world – from Hong Kong to Saudi Arabia, from Vancouver to Taiwan – UNIFIL–Migrante is able to compile statistics and experiences which highlight the disparities between monies collected and spent by the Philippine government for overseas workers' welfare. While the issue of fees is a national issue, it is through the networking of Filipino NGOs around the world that the provision of welfare services for migrant populations takes on global resonance.

The third campaign relates to minimum wages, and while this too has been an ongoing campaign, what follows pertains to the most recent controversy that resulted in a minimum wage cut of 5 per cent in early 1999. In August 1998, Hong Kong's Provisional Urban Council put forward a proposal to cut the wages of domestic workers by 20 per cent. It was suggested that these measures were a response to the local economic downturn, which in turn was a result of the Asian financial crisis. The wage cut issue was supported by the Employers of Overseas Domestic Helpers Association, a network of vocal employers who advocated further cuts up to a total of 35 per cent (Asia Pacific Mission for Migrant Filipinos (APMMF) 1998: 4). Although the Asian financial crisis was stated as the impetus for these proposals, it was also suggested that migrant worker remittances had increased due to higher exchange rates. There was a lively debate about the issue in the pages of Hong Kong newspapers, and reactions to the wage issue were heated. Some claimed that domestic workers were lucky to get the minimum wage at all, and that if they were unhappy with the cuts were 'most welcome to find greener pastures elsewhere – Singapore, Malaysia, Taiwan' (n.a.s. 1999). Others, however, such as Dennis Miller of City University, suggested that the wage cuts 'reflect very badly on a society that, at the first sight of economic problems, demands pay cuts from the weakest and worst-paid section of the community' (APMMF 1998: 5).

In response to the proposal, a newly formed group called the Asian Migrant Coordinating Body (AMCB) met with the relevant government officials. The AMCB is composed of UNIFIL, the Far East Overseas Nepalese Association (FEONA), the Association of Sri Lankans (ASL), Friends of Thai Hong Kong (FOT–HK) and the Indonesian Group. This was not the first time these migrant organisations had joined forces: in May 1998 they had collectively gone to their respective consulates in order to deliver a joint petition on the protection of migrant worker rights, including the ratification of the UN migrant worker rights convention and the provision of support services for migrants (Varona 1998). The financial crisis played an important role in bringing these organisations together, since their status as labour-sending countries to Hong Kong brought to light how migrant workers were collectively being affected by the crisis. In the Philippines, the government was proposing increased fees from overseas workers; in Thailand, escalating unemployment induced the government to provide loans for nationals to seek jobs abroad; and in Indonesia, the government proposed to increase its workers in Hong Kong from 22,000 to 70,000 (Varona 1998). While Nepal and Sri Lanka were not directly hit by the crisis, the flow-on effects for South Asia became apparent for migrant labour. While the interests of each organisation differ in terms of issues needing to be addressed – from government policy in the Philippines and Indonesia to issues of unequal pay for Sri Lankans in Hong Kong – it was clear that the crisis was adversely affecting domestic workers of all nationalities.

The AMCB was not successful in obstructing cuts to domestic workers' minimum wage, but their efforts saw the amount of the cut reduced to 5 per cent instead of 20 per cent. In the AMCB's statement on the outcome of deliberations, the following issues were put forward:

> The salary of foreign domestic helpers [FDHs] is the lowest for foreign workers in Hong Kong. It has not been increased for the past two years despite the increase in the rate of inflation in Hong Kong and in the home countries of the FDHs. Even with the minimum wage for FDHs, a large number of Indonesians, Sri Lankans, Nepalis and other nationalities are receiving wages below the minimum wage. Most of them are paid $2,000 or less. The decision to lower the minimum wage will further reduce the actual wage of many FDHs. It is the responsibility of any government that in times of crisis, the wages of foreign and local workers should be protected. The recent pronouncement by the Hong Kong government speaks to the contrary … It is a mockery of universally accepted norms of equality and is contrary to UN and ILO Conventions on the protection of migrant workers … We are one with the local workers in our struggle against any attack on our wage.
>
> (AMCB 1999: n.p.)

Here again NGOs articulate global norms of equality, but their solidarity with 'local workers' is also stressed. The wage issue in Hong Kong has facilitated links between migrant organisations and the Hong Kong Council of Trade Unions (HKCTU), who supported the wage campaign and participated in the February 1999 protest action against the 5 per cent wage cut. What is particularly interesting about the formation of the AMCB, however, is that it is the first time NGOs from different Asian countries have come together on the issue of domestic work. Previously these organisations had worked independently, since there were practical obstacles such as language, culture and histories/politics of domestic work to work out. That this coalition did come into being reflects the important perspective in Hong Kong of domestic workers as 'workers', and attempts to build solidarity on this basis. It also reflects the financial imperatives of domestic workers themselves, who migrate to Hong Kong primarily for economic reasons.

Conceiving domestic workers as 'workers' – rather than, for example, Filipino women – shifts the debate for Filipino activists in Hong Kong. The perspective of the Filipino feminist–nationalist movement, which constructs the female migrant worker subject as dependent upon neocolonial government policies and negative images of Filipino women abroad, is decentred by Indonesian or Nepali women in Hong Kong who earn below the minimum wage. While it may not be possible to integrate these issues into Philippine national campaigns, by drawing links between labour-sending countries and the impact of the financial crisis on migrant women the AMCB is able to constitute a new domestic worker subject. This subject, however, loses the specificity of national debates.

# Ethnoscapes of activism

> The globalisation of labour and commodity markets means that not only bodies and commodities but also symbols, signs, labels and representations circulate in transnational space.
>
> (Mackie 1998: 45)

What the three campaigns described above illustrate is the importance of NGO advocacy work in a variety of contexts, and how events in Hong Kong constitute and are themselves constituted by events occurring as close as Manila, but as far away as Singapore or Beijing. In terms of the minimum wage issue, the economic and political situations in countries such as Indonesia or Thailand also become important, since the Indonesian government's decision to send more domestic workers to Hong Kong was just as important as the deflated value of the Thai baht in facilitating the coalition of the AMCB. Filipino NGOs make use of the global flow of information on issues regarding the inequalities of transnational labour migration – from immigration policy to government fees to fair wages – and how to use this information to address labour migration from the Philippines to Hong Kong in terms of the New Conditions of Stay, Authentification Fees and minimum wages. It is in this sense that NGOs are operating in transforming terrains which expand the discursive field of their activities, which in turn might affect global conceptions of culturally sensitive issues such as human rights. While their work highlights the importance of Philippine national debates, their activities also have global – or at least regional – resonances. The transforming terrains which connect Manila to Hong Kong, but also to Beijing, Bangkok, Jakarta, Singapore and so on, bear resemblance to Appadurai's notion of 'ethnoscapes' and have implications for understanding transformations in Asian cultural politics.

One of the inspirations for Appadurai's (1996) analysis is the dilemma for anthropology, where the study of 'local culture' is no longer possible in a global and deterritorialised world. The activities of Filipino NGOs are a good example of this contemporary anthropological predilection, since the dynamics of their activities are increasingly subject to forces beyond the Philippines. The NGOs described here, for example, work locally in Hong Kong but network with local, national and global organisations as well as lobby the Hong Kong and Philippine governments. Their perspectives reflect the feminist–nationalist movement in the Philippines, but also the ethics of Hong Kong trade unions and United Nations conventions. Networks with Thai, Indonesian, Sri Lankan and Nepali organisations further demonstrate the complexity of the politics of domestic work in terms of broadening their analyses to include other nationals within discussions of domestic work in Hong Kong. Understanding the networks of NGOs therefore requires new kinds of conceptual strategies for comprehending what otherwise might be understood as 'local activism'. As Smith (1994: 25) has suggested: 'the type of "grassroots" political practice that has emerged among transnational migrants and refugees does not fit well into the restrictive boundaries of local politics conventionally used in connecting the local to the global'.

How to conceptualise such activism or labour migration itself, however, is less apparent. There is a growing literature which examines the networks produced through globalisation generally, and through transnational migration in particular, which is worth detailing here. These studies suggest the importance of theorising networks to better represent new forms of social practice and political space that are now subject to the dynamics of the global. Smith (1994), for example, has argued that the common conceptual strategy of 'it all comes together in LA' – as used by Soja (1989) and other writers in the 'world cities' literature – disregards some of the networks produced through mobile populations as a result of focusing on one particular place. More specifically, Smith argues that:

> The localising move implicit in the world cities problematic obscures the ways that the networks and circuits in which transnational migrants and refugees are implicated constitute fluidly bounded transnational or glob-alised social spaces in which new transnational forms of political organisation, mobilisation, and practice are coming into being.
>
> (Smith 1994: 15; see also Wilson and Dissanayake 1996)

In short, it is not enough to localise the transnational advocacy of NGOs in Hong Kong, contextualising it as a contemporary form of local politics. Rather, it is important to consider Hong Kong as one 'site' of transnational activism within a broader 'social space' where new alliances between migrant, feminist and workers' organisations are taking place.

The globalising social spaces to which Smith (1994) refers bear resemblance to Keck and Sikkink's (1998) work on global social movements. Keck and Sikkink (1998: 3, 46) argue that transnational advocacy networks are 'a set of relevant organisations working internationally with shared values, a common discourse, and dense exchanges of information', but 'must also be understood as political spaces, in which differently situated actors negotiate – formally or informally – the social, cultural, and political meanings of their joint enter-prise'. The Migrant Caucus at the conference in Beijing is an obvious example, since NGOs came together from around the world and debated the situation of migrant women. In sharing the experiences of migrant workers, the language of 'rights' created a political opening for discussions about the discriminating tendencies of employment contracts and immigration policy. Networks such as the AMCB build on this experience and mobilise a conception of migrant women as 'workers': that is, the shared values and common discourse in this case are based on economic identities in the Asian region. This is not surprising given the pattern of poorer women migrating to wealthier economic 'tigers' such as Hong Kong to engage in domestic work, and given the practice of paying domestic workers from countries other than the Philippines lower wages. Furthermore, the formation of the AMCB was motivated by the finan-cial crisis and its aftermath, and in response to minimum wage issues. The political space created by the AMCB therefore challenges unequal geopolitical and economic relations in the Asian region, highlighting how transnational migration is thoroughly entwined with economics.

The globalising social/political space produced by NGO activism expands the discursive field of Filipino NGOs to include other nations alongside the Philippines as unequal players in the global economy. Here, different actors negotiate the effects of transnational migration and strategies to bring about social and political change. Whether or not the human rights discourse mobilised by Filipino NGOs will be the common discourse of the AMCB will depend on the extent to which Indonesian or Sri Lankan organisations wish to engage in the very sensitive arena of political activism in Hong Kong. There are a growing number of domestic workers from countries such as Indonesia and Nepal precisely because they work for lower wages, and because a lack of political organisation within these communities means domestic workers have less recourse to assistance with their individual situations. Filipino NGOs, in contrast, are very vocal and stage protests in central Hong Kong on a regular basis. Indeed, some Filipino domestic workers are now wary of NGO activities, since they are seen to be jeopardising their future employment. In any case, it will be interesting to see whether human rights discourse gains transcultural resonance for the multicountry AMCB, given the emphasis on 'Asian values' and the much vaunted cultural specificities of human rights. For Filipino organisations, global social movements can help bolster their national agendas, but the stability of the AMCB might be jeopardised because of the importance of national politics in labour migration.

## Conclusion

Transnational activism is not a new global phenomenon. Keck and Sikkink (1998) have discussed some of the historical precursors to transnational advocacy networks such as women's suffrage in the United States; antislavery in the US, Britain and its colonies; footbinding in China; and female circumcision in Kenya. What is new about the networks described here, however, is their widespread appearance around the world in the form of 'diasporic public spheres' or arenas of action where the nation is 'no longer the key arbiter of important social change' (Appadurai 1996: 4). In Hong Kong, as elsewhere, new forms of politics and new kinds of political spaces are opening up in response to transnational populations.

Appadurai (1996: 158) has asked the question of what it means to think 'post-nationally'. If diasporic public spheres are part of a post-national imaginary, then the activities of NGOs in Hong Kong represent one example of how the 'perspectival constructs' of Filipino migrant organisations are shifting as a result of expanding political spheres. NGOs are creating new political spaces that reflect the complex circumstances of transnational migration. Their advocacy work reflects the importance of nation in migration – that is, the importance of specific political contexts and cultures in mobilising political action – as well as new forms of post-national solidarity with other Asian countries. There is a need to theorise what Smith (1994: 31) calls a 'politics of simultaneity' or a politics that brings together multiple actors from multiple places. Here I have described a landscape of labour migration that brings Filipinos together with people from places as diverse as Hong Kong, Thailand and Sri Lanka, and the tensions between national and transnational politics.

While the economic consequences of the financial crisis in Asia have been well publicised and documented, it is important to examine how the crisis is also affecting the social and political spaces in Asia. As governments revoke their policies on importing guest workers or propose cuts to the wages of those workers, or as they introduce new fees for overseas workers, they are setting new political agendas. In this chapter I have outlined grassroots resistance to these agendas in Hong Kong, how activism is informed by transnational movements and how the economic crisis is giving rise to new coalitions of migrant organisations. These networks highlight how regional economic and political events can simultaneously create new regional alliances between different nations, shifting the terrain for cultural politics.

## Acknowledgements

The University of Western Sydney, Nepean, provided funding for this research. I would also like to thank the following organisations for their assistance: Mission for Filipino Migrant Workers; Asia-Pacific Mission for Migrant Filipinos; Asian Migrant Center; Asian Domestic Workers Union; Philippine Domestic Workers Union; The Hong Kong Bayanihan Trust; and the Philippine Pastoral Ministry.

## Notes

1   The Overseas Employment Program was the first government initiative to encourage Filipino men and women to work on contracts overseas. President Marcos adopted this strategy in 1972 to 'ease the growing problem of unemployment, in a period of rapid population increase and declining real incomes, and as a means of gaining foreign exchange, through remittances, in the face of recurring balance of payment deficits' (May and Mathews 1997: iii). Employment contracts were first established in the Middle East, and later in the growing economies of Asia, as I discuss in the chapter.
2   On ethnic and gender discrimination see David (1991), Lane (1992), Tyner (1996), Constable (1997), Mackie (1998). On domestic stress see Osteria (1997).
3   As at 28 June 1999, this rate was HK$1 = US$0.128915.
4   It is difficult to obtain current statistics, but in 1993 there were approximately 101,000 Filipino domestic workers, 7,000 Thai domestic workers, 5,000 Indonesian domestic workers and smaller numbers of Sri Lankan, Indian, Malaysian, Burmese, Nepalese and Vietnamese domestic workers (Constable 1997: 3). The figures for Thai and Indonesian workers have increased substantially since this time, and Indonesia alone now sends about 22,000 domestic workers to Hong Kong (Varona 1998).
5   GABRIELA is an umbrella organisation for grassroots women's organisations in the Philippines. Migrante International is also an umbrella organisation, but specifically for NGOs working with migrant labourers.

# References

Aguilar, F. (1996) 'The dialectics of transnational shame and national identity', *Philippine Sociological Review* 44, 1–4: 101–36.

Appadurai, A. (1996) *Modernity at Large: Cultural Dimensions of Globalisation*, Minneapolis: University of Minnesota Press.

Asia Pacific Mission for Migrant Filipinos (APMMF) (1998) 'FDH wage cut: a case of moral bankruptcy', *News Digest* July–December 1998: 4–6.

Asian Migrant Centre (AMC) (1995) 'Statement of the NGO Migrant Caucus to the 4th World Conference on Women', *Asian Migrant Forum* 10: 44–5.

Asian Migrant Coordinating Body (AMCB) (1999) 'We strongly denounce the decision of the Hong Kong government to lower the minimum wage of foreign domestic workers!' Online. Available http://www.hk.super.net/migrant (5 January 1999).

Battistella, G. (1993) 'Networking of NGOs for migrants and information in Asia', *Asian Migrant* VI, 1: 25–31.

Buckley, S. (1997) 'The foreign devil returns: packaging sexual practice and risk in contemporary Japan', in L. Manderson and M. Jolly (eds) *Sites of Desire/Economies of Pleasure: Sexualities in Asia and the Pacific*, Chicago: University of Chicago Press.

Constable, N. (1997) *Maid to Order in Hong Kong: Stories of Filipina Workers*, Ithaca and London: Cornell University Press.

David, R.D. (1991) 'Filipino workers in Japan: vulnerability and survival', *Kasarinlan* 6, 3: 9–23.

de Guzman, A. (1994) 'Export of Filipino labour: problems and prospects for change', *Solidarity* 141: 59–65.

Diaz, C.J. (1993) 'The role of NGOs in the protection and promotion of human rights in Asia', *Asian and Pacific Migration Journal* 2, 2: 199–222.

Gonzalez, J.L., III (1998) *Philippine Labour Migration: Critical Dimensions of Public Policy*, Singapore: Institute of Southeast Asian Studies.

Holt, E. (1996) 'Writing Filipina–Australian bodies: the discourse on Filipina brides', *Philippine Sociological Review* 44, 1–4: 58–78.

Huang, S. and Yeoh, B.S.A. (1996) 'Ties that bind: state policy and migrant female domestic helpers in Singapore', *Geoforum* 27, 4: 479–93.

Keck, M.E. and Sikkink, K. (1998) *Activists Beyond Borders: Advocacy Networks in International Politics*, Ithaca and London: Cornell University Press.

Lane, B. (1992) 'Filipino domestic workers in Hong Kong', *Asian Migrant* V, 1: 24–32.

Lowe, C. (1997) 'The outsider's voice: discourse and identity among the Filipino domestic workers in Hong Kong', paper presented at the *International Conference on Gender and Development in Asia*, 27–9 November, Chinese University of Hong Kong, Hong Kong SAR, China.

Mackie, V. (1994) 'Female labour migration to Japan: the politics of representing women workers', paper presented to the *Linking Our Histories: Asian and Pacific Women as Migrants Conference*, 30 September to 2 October, Melbourne, Australia.

——(1998) ' "Japayuki Cinderella girl": containing the immigrant other', *Japanese Studies* 18, 1: 45–63.

Margold, J.A. (1995) 'Narratives of masculinity and transnational migration: Filipino workers in the Middle East', in A. Ong and M.G. Peletz (eds) *Bewitching Women, Pious Men: Gender and Body Politics in South-East Asia*, Berkeley, CA: University of California Press.

May, R.J. (1997) 'The domestic in foreign policy: the Flor Contemplacion case and Philippine–Singapore relations', *Pilipinas* 29: 63–76.

May, R.J. and Mathews, P. (1997) 'Forward', *Pilipinas* 29: iii–vi.

Mission for Filipino Migrant Workers (MFMW) (1996) 'Human rights and the new conditions of stay'. Online. Available http://www.hk.super.net/migrant/ hrmig (30 March 1999).

name and address supplied (n.a.s.) (1999) 'Pay according to maid's skills', *South China Morning Post*, Hong Kong, 1 March.

Osteria, T.S. (1997) 'Gender dimensions in Philippine overseas migration: impact on family gender power relations', paper presented at the *International Conference on Women in the Asia-Pacific Region: Persons, Power and Politics*, 11–13 August, National University of Singapore, Singapore.

Paz Cruz, V. and Paganoni, A. (1989) *Filipinas in Migration: Big Bills and Small Change*, Manila: Scalabrini Migration Centre.

Pratt, G. (1997) 'Stereotypes and ambivalence: the construction of domestic workers in Vancouver, British Columbia', *Gender, Place and Culture* 4, 2: 159–77.

Rafael, V.L. (1997) ' "Your grief is our gossip": overseas Filipinos and other spectral presences', *Public Culture* 9: 267–91.

Robinson, K. (1996) 'Of mail order brides and "boys' own" tales: representations of Asian–Australian marriages', *Feminist Review* 52: 53–68.

Santos, M.A. (1989) 'An alternative way of looking at women and migration: why an NGO perspective', in *The Trade in Domestic Helpers: Causes, Mechanisms and Consequences*, selected papers from the Planning Meeting on International Migration and Development, Asian and Pacific Development Centre, Kuala Lumpur, Malaysia.

Smith, M.P. (1994) 'Can you imagine? Transnational migration and the globalisation of grassroots politics', *Social Text* 39: 15–33.

Soja, E. (1989) *Postmodern Geographies: The Reassertion of Space in Critical Social Theory*, London: Verso.

Tellez, C.C.A. (1993) 'The protection of the rights of migrant workers in Hong Kong', in G. Battistella (ed.) *Human Rights of Migrant Workers: Agenda for NGOs*, Quezon City: Scalabrini Migrant Center.

Tolentino, R.B. (1996) 'Bodies, letters, catalogues: Filipinas in transnational space', *Social Text* 14, 3: 49–76.

Tyner, J.A. (1996) 'Constructions of Filipina migrant entertainers', *Gender, Place and Culture* 3, 1: 77–93.

United Filipinos in Hong Kong (UNIFIL) (1999) 'Increase in fees for OCWs! The Estrada Government's answer to lowering the minimum wage'. Online. Available http://www.hk.super.net/migrant/unif0399 (accessed 3 April 1999).

United Filipinos in Hong Kong (UNIFIL)–Migrante (1996) 'Position paper on MOI#08'. Online. Available http://www.hk.super.net//migrant/moiposi.htm (accessed 3 April 1999).

Varona, R. (1998) 'A year after: surveying the impact of the Asian crisis on migrant workers'. Online. Available http://is7.pacific.net.hk//amc/papers/crisisdoc4.htm (accessed 20 June 1999).

Wilson, R. and Dissanayake, W. (1996) 'Introduction: tracking the global/local', in R. Wilson and W. Dissanayake (eds) *Global/Local: Cultural Production and the Transnational Imaginary*, Durham and London: Duke University Press.

Yeoh, B.S.A. and Huang, S. (1999) 'Spaces at the margins: migrant domestic workers and the development of civil society in Singapore', *Environment and Planning A* 31: 1149–67.

# Index

Page references for tables and figures are in *italics*; those for notes are followed by n

230    *Index*